Quantum Field Theory in Curved Spacetime and Black Hole Thermodynamics

Chicago Lectures in Physics

Robert M. Wald, Editor
Henry J. Frisch
Gene F. Mazenko
Sidney R. Nagel

Quantum Field Theory in Curved Spacetime and Black Hole Thermodynamics

Robert M. Wald

The University of Chicago Press
Chicago and London

The University of Chicago Press, Chicago 60637
The University of Chicago Press, Ltd., London

Printed in the United States of America
12 11 10 7 8 9

ISBN: 0-226-87025-1 (cloth)
0-226-87027-8 (paper)

Library of Congress Cataloging-in-Publication Data

Wald, Robert M.
Quantum field theory in curved spacetime and black hole thermodynamics / Robert M. Wald.
p. cm. -- (Chicago lectures in physics)
Includes bibliographical references and index.
1. Quantum field theory. 2. Black holes (Astronomy) 3. Gravitational fields. 4. Space and time. 5. Thermodynamics.
I. Title. II. Series.
QC174.45.W35 1994
530.1'43--dc20 94-11065
CIP

The paper used in this publication meets the minimum requirements of the American National Standard for Information Sciences—Permanence of Paper for Printed Library Materials, ANSI Z39.48-1992.

To Barbara, Sarah, and Kristina

Contents

Preface

The subject of quantum field theory in curved spacetime is the study of the behavior of quantum fields propagating in a classical gravitational field. It is used to analyze phenomena where the quantum nature of fields and the effects of gravitation are both important, but where the quantum nature of gravity itself is assumed not to play a crucial role, so that gravitation can be described by a classical, curved spacetime, as in the framework of general relativity. Its two applications of greatest interest are to phenomena occurring in the very early universe and to phenomena occurring in the vicinity of black holes. During the past twenty-five years such phenomena have been explored theoretically, and some unexpected and intriguing results have been obtained. Most prominent among these was the discovery by Hawking that particle creation occurs in the vicinity of black holes. As a direct consequence, a deep connection was obtained between the laws of black hole physics and the ordinary laws of thermodynamics. The Hawking effect and its implications are probably the most valuable clues we have, at present, as to the fundamental features that a quantum theory of gravity is likely to possess.

A student or researcher wishing to learn these developments faces a number of impediments. In my view, the most serious among these is the fact that the standard treatments of quantum field theory in flat spacetime rely heavily on Poincaré symmetry (usually entering the analysis implicitly via plane-wave expansions) and interpret the theory primarily in terms of a notion of "particles". Neither Poincaré (or other) symmetry nor a useful notion of "particles" exists in a general, curved spacetime, so a number of the familiar tools and concepts of field theory must be "unlearned" in order to have a clear grasp of quantum field theory in curved spacetime.

One of the main goals of these lecture notes is to provide a coherent, pedagogical introduction to the formulation of quantum field theory in curved spacetime. We begin with a treatment of the ordinary, one-dimensional, quantum harmonic oscillator and logically progress, through the construction of quantum field theory in flat spacetime, to the possible constructions of quantum field theory in curved spacetime, and, ultimately, to the algebraic formulation of the theory. It is my hope that this presentation will help disentangle the essential features of the theory from some inessential ones (such as a particle interpretation), and that it will help clarify the relationships between the various different approaches that have been taken toward the formulation of the theory.

The other main goal of these lecture notes is to provide a comprehensive, up-to-date account of the Unruh effect, the Hawking effect, and some of the ramifications of the Hawking effect, particularly with regard to black hole thermodynamics. Although the key results on these topics were obtained in the 1970's, they remain an active area of current research.

The reader should be made aware that it has *not* been my goal to present a balanced, comprehensive review of all research on quantum field theory in curved spacetime. In particular, it should be noted that, with the exception of some brief discussion of quantum field effects in deSitter spacetime (see section 5.3), there is very little discussion relating to the applications of the theory to cosmology.

In writing these notes, I have assumed that the reader has a knowledge of classical general relativity comparable to the level of a one-term, graduate, introductory course. This amounts to assuming that the reader is comfortable with the notion of spacetime as a manifold together with a Lorentz signature metric, and is familiar with such concepts as tensor fields, covariant derivatives, and geodesics. An introduction to the needed "advanced topics" of global hyperbolicity, Killing horizons, and black holes is given in these notes in a basically self-contained manner.

I also have assumed that the reader has a knowledge of quantum field theory comparable to the level of an introductory graduate course. In principle, only a basic knowledge of ordinary quantum mechanics actually should be necessary, since quantum field theory in flat spacetime is developed *ab initio* in these notes. Indeed, for some of the discussion, it may be more useful for the reader *not* to have had a prior knowledge of standard presentations of quantum field theory. However, readers who are entirely unfamiliar with concepts such as Fock space, annihilation and creation operators, etc., may have a relatively difficult time reading portions of these notes.

I wish to thank Steve Fulling and Ted Jacobson for reading the entire manuscript with great care, and making many valuable suggestions for improvements. Similar thanks are due Rainier Verch for reading chapter 4, and I also wish to thank Bernard Kay for a number of helpful comments. I am greatly indebted to Miles Blencowe for carefully proofreading the entire manuscript and suggesting some improvements in the exposition. Chapters 2, 3, and 4 of these notes as well as parts of chapters 5 and 7 correspond closely to my lectures presented at the 1992 Les Houches School on Gravitation and Quantization. I wish to thank Elsevier Science Publishers for waiving any time restrictions on my use of this material from my Les Houches lectures. Support from NSF grant PHY-9220644 to the University of Chicago also is gratefully acknowledged.

Notation, Conventions, and Terminology

Our notation and conventions for spacetime quantities follow Wald (1984a). In particular, we choose metric signature -+++ and employ the abstract index notation for tensor fields on spacetime used in that reference. In a few places (particularly in sections 3.3 and 4.4) we also shall employ a similar abstract index notation for elements of the tensor products of Hilbert spaces. The rules for this index notation are explained in Appendix A.3. Since our use of the index notation for tensors over Hilbert spaces is quite limited and takes place in a context where no confusion with spacetime tensors should occur, we shall use lower case latin letters for the abstract indices of tensors in both the Hilbert space and spacetime cases.

A number of special symbols defined in the book are used frequently but are not always redefined whenever they appear. For the convenience of the reader, a notation index of these symbols—which provides a brief definition of each symbol as well as the page number on which it is defined—can be found at the end of the book.

One point of terminology should be mentioned here. The term "vacuum state" has a number of independent connotations. One such connotation is that of being a ground state (i.e., a state of minimum energy) with respect to some notion of "time translations". The standard vacuum state of free quantum field theory in Minkowski spacetime has this property, and the "vacuum states" for stationary spacetimes that will be constructed in section 4.3 also are ground states. However, in a general, nonstationary, curved spacetime there is no preferred notion of "time translations", and, in general, no useful notion of a "ground state" exists. Another connotation of the term "vacuum state" (for the quantum field theory of a linear field) is one where the correlation functions of the field correspond to having "Gaussian fluctuations"; equivalently, all the "truncated n-point functions" of the field vanish. This notion makes sense in a general, curved spacetime; in the algebraic approach, it corresponds to the notion of a "quasi-free state" (see section 4.5). Yet a third connotation of "vacuum state" is that of a particular vector in a Fock space constructed from a one-particle Hilbert space comprised by a suitably chosen complex subspace of solutions to the classical field equations. Such "vacuum states" automatically have vanishing truncated n-point functions (i.e., they are automatically quasi-free) and, in addition, are pure states in the sense defined in section 4.5; conversely, any pure quasi-free state arises as the "vacuum vector" in a suitable Fock space construction. This third notion of a "vacuum state" is the one used by most researchers in quantum field theory in curved spacetime, and I shall employ this terminology here. Thus, in these notes, the term "vacuum state" means, in the algebraic language, a "pure, quasi-free state".

1 Introduction and Overview

The main goals of these notes are (i) to present a mathematically precise formulation of the theory of quantum fields in curved spacetime, (ii) to present some of the principal results of the theory—particularly the Unruh and Hawking effects—and (iii) to discuss, in depth, some of the ramifications of the Hawking effect, particularly with regard to the remarkable analogy between the laws of black hole physics and the laws of thermodynamics. In this chapter, we give a brief introduction to these ideas and give an overview of the contents of these notes.

1.1 The Formulation of Quantum Field Theory in Curved Spacetime

In quantum field theory in curved spacetime, one treats gravitation classically, as in the framework of general relativity. Thus, spacetime structure is described by a manifold, M, on which is defined a classical, Lorentz signature metric, g_{ab}. One thereby avoids confronting the fundamental difficulty of how to formulate quantum field theory without a classical background metrical (and causal) structure of spacetime. The matter fields propagating in this classical spacetime are then treated as quantum fields, in the same sort of manner as in the theory of a charged scalar (or Dirac) field propagating in an external electromagnetic field. As we shall see, for linear fields in curved spacetime, a fully satisfactory, mathematically rigorous theory can be constructed.

One expects that quantum field theory in curved spacetime should have only a limited range of validity. In particular, it certainly should break down—and be replaced by a quantum theory of gravitation coupled to matter—when the spacetime curvature approaches Planck scales. Of course, the precise criteria for the validity of quantum field theory in curved spacetime will be known only when the ultimate theory of quantum gravity is available. Nevertheless, one would expect that its range of validity should include a wide variety of phenomena of interest, such as particle creation near a black hole whose Schwarzschild radius is much greater than the Planck length.

The difficulties involved in explaining the formulation of quantum field theory in curved spacetime to a reader familiar with standard treatments of quantum field theory in flat spacetime is somewhat analogous to the difficulties involved in explaining general relativity to a reader familiar with special relativity in the manner in which it normally is formulated—where primary emphasis is placed upon the existence of global families of inertial observers and the relationships between these families described by Poincaré transformations. Neither the notion of global inertial observers nor Poincaré transformations generalize in a meaningful way to curved spacetime. However, when one recognizes that the structure of spacetime in special relativity is most naturally and simply described by a flat spacetime metric—and that the existence of global families of inertial observers may be viewed as a secondary consequence of the presence of this flat metric—the transition to the framework of general relativity is straightforward: One simply allows the spacetime metric to be curved.

In a similar manner, in quantum field theory in flat spacetime, the Poincaré group plays a key role in picking out a preferred vacuum state and defining the notion of a "particle". In the past, much attention has been devoted to the issue of how to generalize the notion of "particles" to curved spacetime. One of the key points which will be emphasized by our presentation here is that this issue is irrelevant to the formulation of quantum field theory in curved spacetime—in much the same manner as the issue of how to generalize the definition of global inertial coordinates to curved spacetime is irrelevant to the formulation of general relativity. Quantum field theory is a quantum theory of *fields,* not particles. Although in appropriate circumstances a particle interpretation of the theory may be available, the notion of "particles" plays no fundamental role either in the formulation or interpretation of the theory.

In these notes, we shall present quantum field theory directly as a theory resulting from applying the principles of quantum theory to a classical field system. We thereby shall circumvent the usual convoluted historical and pedagogical route toward the construction of quantum field theory, wherein one first pretends that one is constructing the theory of a relativistic particle, then recognizes difficulties with that theory, and then cures these difficulties via "second quantization." In this manner, it will be clear from the outset that the fundamental observables of the theory are field amplitudes and momenta. As we shall see, in flat spacetime—and, more generally, in a curved stationary spacetime—a natural particle interpretation will emerge when we couple the field to a simple model quantum mechanical system (i.e., a "particle detector") and investigate the effects of the interaction. In curved spacetimes which are asymptotically stationary in the past and/or future, natural particle interpretations are also available in these asymptotic regions. However, in other circumstances, the notion of "particle" is, at best, of very limited utility.

The quantum theory of a field system differs from the quantum theory of a particle system primarily in that a field system has infinitely many degrees of freedom. Now, for a system with finitely many degrees of freedom, the "kinematical structure" of the theory is entirely determined by the canonical commutation relations for the position and momentum operators. More precisely, as we shall review in section 2.2, the Stone-von Neumann theorem implies that these canonical commutation relations determine uniquely (up to unitary equivalence) a choice of Hilbert space $\mathfrak{F}$ and a choice of self-adjoint operators on $\mathfrak{F}$ corresponding to the position and momentum observables. However, the Stone-von Neumann theorem does not hold for a field system, which has infinitely many degrees of freedom. As we shall see explicitly in section 4.4, infinitely many unitarily inequivalent, irreducible representations of the canonical commutation relations exist in quantum field theory.

In flat spacetime, Poincaré symmetry is used to pick out a preferred representation. This selection of a preferred representation is mathematically equivalent to a selection of a preferred "vacuum state", which, in turn, is mathematically equivalent to the selection of a preferred definition of the notion of "particles" in the theory. However, as already indicated above, in a general curved spacetime, there does not appear to be any preferred notion of "particles." Indeed, for a noncompact space, in cases where natural notions of "particles" are available in both the asymptotic past and the asymptotic future, the representations of the canonical commutation relations corresponding to these two notions will, in general, be unitarily inequivalent. (An analogous phenomenon occurs in the "infra-red catastrophe" of quantum electrodynamics.) Thus, it might appear that there exists a fundamental difficulty in formulating quantum field theory in curved spacetime.

This apparent difficulty is cured by formulating the theory via the algebraic approach. In essence, the algebraic approach allows one to consider—on an equal footing—all states arising in all the different (i.e., unitarily inequivalent) Hilbert space constructions of the theory. In this manner, the theory can be formulated in a clear, mathematically precise, manner without requiring at the outset that one pick a preferred representation of the canonical commutation relations, and, in particular, without requiring one to define a preferred notion of "particles".

In our development of quantum field theory in curved spacetime, it is essential that we distinguish between "universal" elements of the theory—which carry over without essential change from flat spacetime to curved spacetime—from those elements of quantum field theory in flat spacetime which rely on Poincaré symmetry to pick out a preferred representation of the canonical commutation relations. These latter elements generalize in a natural way only to very special curved spacetimes (in particular, stationary spacetimes), and they play no essential role when the theory is reformulated via the algebraic approach. In effect, the "plane-wave expansions" used in standard textbook treatments of quantum field theory in flat spacetime mix together these different elements. Hence, in these notes, we will devote considerable attention to formulating quantum field theory in flat spacetime in a way that disentangles these elements.

In our treatment of quantum field theory, we will focus attention on the case of a real, linear, scalar field. However, we will formulate all the basic ideas so that they will carry over without any essential change to all real, linear bosonic fields. The minor modifications needed to treat complex, linear, bosonic fields will be briefly discussed in section 4.7. The modifications needed to formulate the theory of linear fermionic fields consist, in essence, of several key sign changes, and we also shall describe these in section 4.7. However, with the exception of a few remarks, nonlinear (i.e., self-interacting) fields will not be treated in these notes.

Historically, the first key issue to stimulate the development of quantum field theory in curved spacetime was that of particle creation produced by the expansion of the universe near the "big bang". The basic framework and structure of quantum field theory in curved spacetime emerged by the late 1960's from the analyses of this phenomenon by Parker (1969) and others. The development of quantum field theory in curved spacetime then received enormous further impetus from the discovery by Hawking (1975) that black holes radiate as blackbodies due to particle creation. Many investigations related to quantum field theory in curved spacetime were undertaken in the mid-to-late 1970's. (A comprehensive summary of much of this work can be found in the book of Birrell and Davies (1982) and a more up-to-date review with more emphasis on mathematical and foundational issues can be found in Fulling (1989).) Although the level of intensity of effort has declined considerably since then, research on quantum field theory in curved spacetime has continued. In particular, research on some key issues related to the Hawking effect remains active, although some of this work extends beyond the borders of quantum field theory in curved spacetime.

The mathematically rigorous formulation of quantum field theory in curved spacetime is based mainly on the work of Segal (1963, 1967) and others on the general theory of linear dynamical systems and on the algebraic approach to field theory initiated by Haag and Kastler (1964) (see Haag (1992) for a recent comprehensive review). Most of the elements of this body of work which do not make explicit reference to the Poincaré group can be taken over directly to quantum field theory in curved spacetime. Some new issues which arise in quantum field theory in curved spacetime involve the definition of the stress-energy tensor of the quantum field and the specification of a preferred class of states on which the stress-energy is defined. Research on some mathematical questions raised by these issues continues to be pursued.

It is worth noting that most of the available treatments of quantum field theory in curved spacetime either are oriented strongly toward mathematical issues (and deal, e.g., with C^*-algebras, KMS states, etc.) or are oriented toward a concrete physical problem (and deal, e.g., with particular mode function expansions of a quantum field in a certain spacetime). An important goal of these notes is to bridge some of the gap that exists between these two approaches to the subject.

We shall begin in the next chapter with some preliminaries on the quantum theory of particle systems, i.e., systems with finitely many degrees of freedom. (Some mathematical preliminaries on Hilbert spaces are given in the Appendix.) In chapter 3 we shall develop the theory of a free Klein-Gordon scalar quantum field in flat spacetime. The generalization of this construction to globally hyperbolic, curved spacetimes then will be given in chapter 4. In that chapter, we also will analyze in detail the issue of the unitary equivalence of different constructions. This issue is of direct mathematical importance with regard to the formulation of the theory and also is of direct physical importance for the calculation of the S-matrix—describing particle creation and scattering—in situations where asymptotic notions of "particles" exist. The formulation of quantum field theory in curved spacetime via the algebraic framework then will be presented in section 4.5. In section 4.6 we shall give an analysis of the expected stress-energy tensor of a quantum field and discuss how "back-reaction" is treated in the theory. The "Hadamard condition" on states will be introduced in the course of this analysis. Chapter 4 concludes with a brief discussion of how the construction of the quantum theory of a scalar field in curved spacetime can be generalized to other linear boson and fermion fields.

The remainder of these notes will be devoted to applications of quantum field theory in curved spacetime to the analysis of phenomena related to the "Hawking effect" of particle creation by black holes. In chapter 5, we shall explain how it happens that a uniformly accelerating observer in flat spacetime "feels himself" to be immersed in a thermal bath of particles when the quantum field is in its vacuum state. The generalization of this "Unruh effect" to curved spacetime also is presented in chapter 5. Chapter 6 lays the groundwork for a discussion of the Hawking effect by presenting some key results in the classical theory of black holes, particularly the laws of black hole mechanics. Finally, in chapter 7 we present a derivation of the Hawking effect and discuss its ramifications.

We now give a brief introduction to the key issues and ideas contained in chapters 5, 6, and 7.

1.2 Classical Black Hole Thermodynamics

In physical terms, a "black hole" is a region where gravity is so strong that nothing—not even light—can escape. A mathematically precise definition of a black hole can be given for a spacetime which is asymptotically flat: Any events in the spacetime which do not lie to the past of the family of all observers who remain in the (nearly flat) asymptotic region for all time are said to be contained within a black hole. The boundary of a black hole is called its event horizon.

A theorem of Schoen and Yau (1983) asserts that a trapped surface must form whenever a sufficiently large quantity of matter is compacted into a sufficiently small region. It then follows from the singularity theorems of classical general relativity (see, e.g., Hawking and Ellis (1973)) that a spacetime singularity must occur. As we shall discuss further in section 6.1, the cosmic censorship conjecture asserts that any singularity produced in this manner must be hidden within a black hole, i.e., the complete gravitational collapse of a star or other body will always result in a black hole rather than a "naked singularity". Although there remains little direct evidence concerning the validity of this conjecture within classical general relativity, most researchers believe that it holds, and, hence, that black holes are physically relevant objects, describing the endpoint of gravitational collapse.

The theory of black holes in classical general relativity was largely developed in the late 1960's and early 1970's. A number of remarkable results were obtained. In particular, it was shown that stationary (i.e., "time-independent") black holes with a "non-degenerate" horizon in Einstein-Maxwell theory (i.e., general relativity with no matter fields present except source-free electromagnetic fields) are uniquely determined by only three parameters: mass, angular momentum, and charge. (See, e.g., Wald (1984a) for a summary of these uniqueness theorems.) In addition, it was shown that energy can be extracted from stationary black holes which are rotating or charged. However, perhaps the most remarkable result obtained in these investigations was the discovery of a close mathematical analogy between certain laws of "black hole mechanics" and the ordinary laws of thermodynamics (Bardeen et al, (1973)). In this analogy, the mass of the black hole plays the role of the energy of a thermodynamic system, its surface gravity plays the role of temperature, and the area of its event horizon plays the role of entropy. Chapter 6 is primarily devoted to deriving the laws of black hole mechanics in classical general relativity.

A hint that the analogy between the laws of black hole mechanics and thermodynamics might extend beyond a formal, mathematical similarity is contained in the fact that the quantities appearing in one pair of mathematical analogs—namely, black hole mass and thermodynamic energy—are, in fact, physically analogous: In general relativity, the mass of a black hole is the same quantity as its total energy. However, the physical relationship breaks down beyond this point in classical theory, since, clearly, the physical temperature of a black hole in classical general relativity is absolute zero.

1.3 The Hawking Effect and its Ramifications

The gravitational collapse of a body to form a black hole is a violent event, so it would be expected that in quantum field theory, particle creation should occur in the strongly time-dependent gravitational field associated with the collapse. Particle creation effects occurring at "late times" after the black hole has "settled down" to its stationary final state also would be expected in the case of a rotating or charged black hole, where energy extraction is possible classically. However, it came as a great surprise when an analysis by Hawking (1975) showed that particle creation also occurs at "late times" following the collapse of a body to a nonrotating, uncharged black hole. It came as an even greater surprise that this radiation which, in effect, is "emitted" by the black hole, has a remarkably simple and familiar form: It corresponds to perfectly thermal, "blackbody" radiation, at temperature $kT = \kappa/2\pi$, where κ denotes the surface gravity of the black hole. We shall present the derivation of this result in section 7.1.

The Hawking effect is closely related mathematically to an effect subsequently discovered by Unruh (1976), who found that an accelerating observer in flat spacetime feels himself to be immersed in a thermal bath of particles at temperature $kT = a/2\pi$ (where a denotes the acceleration of the observer) when the quantum field is in its vacuum state as determined by inertial observers. More precisely, the inertial vacuum state of Minkowski spacetime is a thermal ("KMS") state when analyzed with respect to the notion of "time translations" defined by a one-parameter family of Lorentz boosts. (The orbits of these Lorentz boost isometries correspond to a family of uniformly accelerating observers.) This result—known as the "Unruh effect"—is actually a consequence of the fact that the Lorentz boost isometries possess a Killing horizon, i.e., a null surface to which the Killing field generating the isometries is normal. The Unruh effect has a generalization to curved spacetimes possessing a one-parameter group of isometries which have an associated Killing horizon. The close mathematical relationship between the Unruh effect and the Hawking effect stems from the fact that the event horizon of a stationary black hole is a Killing horizon. We shall analyze the Unruh effect and its generalization to curved spacetimes in detail in chapter 5.

An important, immediate consequence of the Hawking effect is that it puts the analogy between the laws of black hole mechanics and the laws of thermodynamics on a firm physical foundation: The surface gravity of a black hole—which plays a role mathematically analogous to temperature—is now seen to be the physical temperature of the black hole in a completely literal sense. Furthermore, the Hawking effect and other closely related quantum effects make viable the "generalized second law of thermodynamics", which asserts that the total entropy of matter outside black holes plus 1/4 the surface area of all black holes never decreases with time. This suggests that the laws of black hole mechanics literally are the ordinary laws of thermodynamics applied to a system containing a black hole. However, the manner in which these laws might arise from a statistical mechanics based upon a complete quantum theory of gravity remains a mystery. The ramifications of the Hawking effect for black hole thermodynamics and the generalized second law will be discussed in detail in section 7.2.

An important further ramification of the Hawking effect concerns an issue posed when "back-reaction" of the quantum field on the black hole is taken into account. Clearly, if total energy is to be conserved, the black hole must lose mass as it emits radiation via the particle creation process. A simple estimate indicates that a black hole should radiate all of its mass—and, thus, "evaporate" completely—within a finite time. Since the black hole emits radiation in a thermal (and, thus, highly mixed) state, it would appear that in the process of black hole formation and evaporation, an initial pure quantum state can evolve to a mixed state. This raises some issues of principle concerning the nature of quantum theory, which presently continue to remain at the center of considerable discussion and debate. The phenomenon of black hole evaporation and the consequent "loss of quantum coherence" will be discussed in section 7.3.

2 Quantum Mechanical Preliminaries

Our formulation of quantum field theory in curved spacetime will be based directly upon the quantum mechanics of systems satisfying linear equations of motion, i.e., harmonic oscillators. Indeed, the only important difference between a linear (i.e., non-self-interacting) field in curved spacetime and a collection of ordinary harmonic oscillators (with possibly time dependent masses and spring constants) is that the field system has infinitely many degrees of freedom. The main goal of this chapter is to reformulate some standard results from the quantum mechanics of harmonic oscillator systems in a manner that will enable us, in the next two chapters, to give a direct formulation of quantum field theory in flat and curved spacetimes. We shall begin with a very brief discussion of the classical mechanics of nonlinear systems with finitely many degrees of freedom, so that we may introduce the symplectic form in a general setting. Then we shall specialize to linear systems, and will write the quantum theory of a collection of harmonic oscillators in a manner that brings to the fore both the role of the symplectic vector space structure of the classical solutions and the freedom available in making a choice of "one-particle Hilbert space". Although the material covered in this chapter is entirely elementary in the sense that the quantum theory we construct is the same as can be found in any introductory text on nonrelativistic quantum mechanics, some aspects of our discussion and notation will be unconventional (within this quantum mechanical setting). It is hoped, thereby, that this chapter will afford a useful link between ideas with which the reader undoubtedly is very familiar and the construction of quantum field theory in curved spacetime given in the subsequent chapters of these notes.

2.1 Classical Dynamics of Particle Systems

The quantum theory of a system composed of finitely many particles (or, more generally, any system with finitely many degrees of freedom) is based directly upon the structure present in the Hamiltonian formulation of the classical theory. In this section, we review the key elements of this classical structure.

The reader undoubtedly is familiar with classical Hamiltonian mechanics in the manner presented in most introductory textbooks. In such treatments, it normally is assumed that the state of a classical system with finitely many degrees of freedom is characterized by n "generalized coordinates", $q_1,\ldots,q_n$, (describing its "configuration") together with n corresponding "conjugate momenta", $p_1,\ldots,p_n$. The collection of all possible values of coordinates and momenta, $(q_1,\ldots,q_n; p_1,\ldots,p_n)$, is referred to as *phase space*. Dynamical evolution on phase space is then determined by specifying a Hamiltonian function, $H = H(q_1,\ldots,q_n; p_1,\ldots,p_n)$, and imposing Hamilton's equations of motion,

$$\frac{dq_\mu}{dt} = \frac{\partial H}{\partial p_\mu}\,, \quad \frac{dp_\mu}{dt} = -\frac{\partial H}{\partial q_\mu} \qquad (2.1.1)$$

We can express these equations more succinctly by writing $y = (q_1,\ldots,q_n; p_1,\ldots,p_n)$ and introducing the antisymmetric $2n\times2n$ matrix $\Omega^{\mu\nu}$, with $\Omega^{\mu\nu} = 1$ when $\nu = \mu + n$ and $\Omega^{\mu\nu} = 0$ when $|\mu-\nu| \neq n$. We obtain, for $\mu = 1,\ldots,2n$,

$$\frac{dy^\mu}{dt} = \sum_{\nu=1}^{2n} \Omega^{\mu\nu} \frac{\partial H}{\partial y^\nu} \qquad (2.1.2)$$

It is not too large a jump to go from Hamilton's equations in the form (2.1.2) to the much more general and abstract formulation of Hamiltonian mechanics given, e.g., in (Arnold 1989), in which the essential mathematical structure which is needed to define the theory can be seen more clearly. In this more abstract formulation, states of a classical system with n degrees of freedom are represented by points in a 2n-dimensional manifold, $\mathcal{M}$, which we continue to refer to as phase space. The fundamental structure on $\mathcal{M}$ needed to formulate Hamiltonian mechanics is a *symplectic form*, Ω_{ab}, by which is meant a nondegenerate, closed, 2-form on $\mathcal{M}$. In other words, Ω_{ab} is a tensor field of type (0,2) on $\mathcal{M}$ such that, $\Omega_{ab} = \Omega_{[ab]}$, $\nabla_{[a}\Omega_{bc]}=0$, (where square brackets denote antisymmetrization) and for any tangent vector v^b on $\mathcal{M}$, we have $\Omega_{ab}v^b = 0$ if and only if $v^b = 0$. The nondegeneracy of Ω_{ab} implies that it has a unique inverse,

The Poisson bracket is manifestly antisymmetric, and it is not difficult to show that it satisfies the Jacobi identity by virtue of the closedness of Ω_{ab}.

Since the canonical coordinates q_μ, p_ν, are, of course, functions on $\mathcal{M}$, they may be viewed as observables. Indeed, these observables are "fundamental" in the sense that all other observables are functions of them. By eqs. (2.1.4) and (2.1.6), the Poisson bracket of these fundamental observables is given by

$$\{q_\mu, q_\nu\} = \{p_\mu, p_\nu\} = 0 \tag{2.1.7}$$

$$\{q_\mu, p_\nu\} = \delta_{\mu\nu} \tag{2.1.8}$$

Our attention in these lectures will be focused almost exclusively on "linear dynamical systems". More precisely, we shall consider systems for which the following two properties hold. (1) We have $\mathcal{M} = T^*(\mathcal{Q})$, where $\mathcal{Q}$ has the natural structure of a vector space. In that case, we can choose a basis of $\mathcal{Q}$, and use the basis components of the vectors in $\mathcal{Q}$ to define "linear coordinates", $(q_1,\ldots,q_n)$, globally on $\mathcal{Q}$. These coordinates then give rise to globally well defined linear canonical coordinates $(q_1,\ldots,q_n; p_1,\ldots,p_n)$ on $\mathcal{M}$ such that eq. (2.1.4) holds. (2) The Hamiltonian, H, is a quadratic function on $\mathcal{M}$, so that the equations of motion, (2.1.1), are linear in the linear canonical coordinates. In that case, the manifold of solutions, $\mathcal{S}$, also acquires a natural vector space structure.

Note that a linear dynamical system is thus simply a collection of coupled harmonic oscillators (possibly with time-dependent and/or negative masses and spring constants). As already mentioned above, a linear field in curved spacetime, is, in essence, an infinite collection of such harmonic oscillators.

A key consequence of property (1) is that it allows us to view the symplectic form, Ω_{ab}, in the following manner. On account of the vector space structure of $\mathcal{M}$, we may identify the tangent space at any point $y \in \mathcal{M}$ with $\mathcal{M}$ itself. Under this identification, the symplectic form, Ω_{ab}, becomes a bilinear function, $\Omega: \mathcal{M} \times \mathcal{M} \to \mathbb{R}$, on $\mathcal{M}$. Furthermore, since, by eq. (2.1.4), the components of Ω_{ab} are constant in the linear canonical coordinate basis, this bilinear map, Ω, is independent of the choice of point y used to make the identification. We shall refer to Ω as a *symplectic structure* on $\mathcal{M}$. For $y_1, y_2 \in \mathcal{M}$, Ω is given explicitly by

$$\Omega(y_1, y_2) = \sum_\mu [p_{1\mu}q_{2\mu} - p_{2\mu}q_{1\mu}] \tag{2.1.9}$$

In this way, the phase space of a linear dynamical system, may be naturally viewed simply as a symplectic vector space, $(\mathcal{M}, \Omega)$—i.e., a vector space, $\mathcal{M}$, on which is defined a nondegenerate, antisymmetric, bilinear map, Ω—rather than as a symplectic manifold, $(\mathcal{M}, \Omega_{ab})$.

For a system satisfying property (1), we may use the symplectic structure, Ω, to, in effect, replace the linear canonical coordinates $(q_1,...,q_n; p_1,...,p_n)$ as follows: Fix $y \in \mathcal{M}$. Then, we may view the quantitiy $\Omega(y,\cdot)$ as a linear function on $\mathcal{M}$. If we choose $y = (0,...,0, q_\mu=1,0,...,0)$, then by eq. (2.1.9) we see that $\Omega(y,\cdot)$ is just the function whose value at an arbitrary point of $\mathcal{M}$ is $-p_\mu$, where "p_μ" is the value of the μth momentum coordinate. Similarly, it is easily checked that for $y = (0,...,0, p_\mu=1,0,...,0)$, we have $\Omega(y,\cdot) = q_\mu$. It follows that the collection of functions of the form $\Omega(y,\cdot)$, with y ranging over $\mathcal{M}$, correspond to arbitrary linear combinations of the linear canonical coordinates $(q_1,...,q_n; p_1,...,p_n)$. Thus, we may rewrite any relations involving the linear canonical coordinates in terms of the functions $\Omega(y,\cdot)$. In particular, the Poisson bracket relations (2.1.7)-(2.1.8) are equivalent to the single relation,

$$\{\Omega(y_1,\cdot), \Omega(y_2,\cdot)\} = -\Omega(y_1, y_2) \tag{2.1.10}$$

It may seem to the reader that eq. (2.1.10) does little more than express the perfectly satisfactory and familiar relations (2.1.7)-(2.1.8) in a much more obscure form. However, even in the finite dimensional case considered here, eq. (2.1.10) has a notable advantage over (2.1.7)-(2.1.8) in that it allows us to write down the fundamental Poisson bracket relations without requiring us to make a particular choice of linear canonical coordinates on $\mathcal{M}$. More importantly, when we treat fields in curved spacetime (for which $\mathcal{M}$ is infinite dimensional), it is far from clear how to generalize the relations (2.1.7)-(2.1.8) in such a way as to define the Poisson bracket on a natural class of linear functions on $\mathcal{M}$. On the other hand, the Poisson bracket relation in the form (2.1.10) generalizes straightforwardly to the infinite dimensional case. Consequently, in these notes, we will work almost exclusively with the functions $\Omega(y,\cdot)$ as the fundamental observables on $\mathcal{M}$, and the reader is urged to become familiar and comfortable with the notation used in eq.(2.1.10).

Only property (1) of linear dynamical systems has been used thus far. We now explore some consequences of property (2), which states that the Hamiltonian function, H, on $\mathfrak{M}$ at any time t takes the form

$$H(t; y) = \frac{1}{2}\sum_{\mu,\nu} K_{\mu\nu}(t) y^\mu y^\nu \tag{2.1.11}$$

where y^μ denotes the linear canonical coordinates, $(q_1,\ldots,q_n; p_1,\ldots,p_n)$, of point y, and where, without loss of generality, we take $K_{\mu\nu} = K_{\nu\mu}$. By eq. (2.1.2), Hamilton's equations of motion are

$$\frac{dy^\mu}{dt} = \sum_{\rho,\nu} \Omega^{\mu\rho} K_{\rho\nu} y^\nu \tag{2.1.12}$$

Now, let $y_1(t)$, $y_2(t)$ be two solutions of the equations of motion (2.1.12) and let

$$\begin{aligned} s(t) &= \Omega(y_1(t), y_2(t)) \\ &= \sum_{\alpha,\beta} \Omega_{\alpha\beta} y_1{}^\alpha y_2{}^\beta \end{aligned} \tag{2.1.13}$$

Then, we have

$$\begin{aligned} \frac{ds}{dt} &= \sum_{\alpha,\beta} \Omega_{\alpha\beta}\left[\frac{dy_1{}^\alpha}{dt} y_2{}^\beta + y_1{}^\alpha \frac{dy_2{}^\beta}{dt}\right] \\ &= \sum_{\alpha,\beta,\rho,\nu} \Omega_{\alpha\beta}\left[\Omega^{\alpha\rho} K_{\rho\nu}\, y_1{}^\nu y_2{}^\beta + \Omega^{\beta\rho} K_{\rho\nu} y_2{}^\nu y_1{}^\alpha\right] \\ &= -\sum_{\beta,\nu} K_{\beta\nu} y_1{}^\nu y_2{}^\beta + \sum_{\alpha,\nu} K_{\alpha\nu} y_2{}^\nu y_1{}^\alpha \\ &= 0 \end{aligned} \tag{2.1.14}$$

Thus, for a linear dynamical system, the "symplectic product", s, of two solutions is conserved. (This result is really a consequence of the much more general fact for nonlinear systems that dynamical evolution defines a canonical transformation on phase space.) Thus, the symplectic structure, $\Omega: \mathfrak{M} \times \mathfrak{M} \rightarrow \mathbb{R}$, on $\mathfrak{M}$ gives rise to a natural symplectic structure (also denoted Ω) on the vector space of solutions, $\mathcal{S}$, since the symplectic structure on $\mathcal{S}$ obtained by identifying $\mathcal{M}$ and $\mathcal{S}$ does not depend upon the choice of "initial time" $t = 0$ under which the identification is made. The explicit form of Ω on $\mathcal{S}$ can be obtained by expressing p_μ as a function of the q's and their time derivatives via Hamilton's equations of motion (2.1.1), and then substituting the result in eq. (2.1.9).

The symplectic vector space structure $(\mathcal{S}, \Omega)$ of the manifold of solutions for a linear dynamical system is the fundamental classical structure that underlies the construction of the quantum theory of a linear field.

2.2 Quantum Mechanics

The mathematical framework of quantum mechanics differs drastically from classical mechanics: A state in quantum theory is represented by a vector in an infinite dimensional Hilbert space, $\mathfrak{F}$, rather than being represented by a point in a finite dimensional manifold $\mathcal{M}$. An observable in quantum theory is represented as a self-adjoint operator acting on $\mathfrak{F}$, rather than being represented by a real-valued function on $\mathcal{M}$. Finally, dynamical evolution is given by a one-parameter family of unitary transformations on $\mathfrak{F}$ generated by a Hamiltonian operator, rather than being given by a one-parameter family of canonical transformations on $\mathcal{M}$ generated by a Hamiltonian function via eq. (2.1.3).

The key issue in the construction of a quantum theory corresponding to a classical system is how to choose the Hilbert space $\mathfrak{F}$ and the self-adjoint operators $\hat{f}_i: \mathfrak{F} \to \mathfrak{F}$ corresponding to the classical observables of interest, f_i. The basic guiding principle in making the choice of $(\mathfrak{F}, \hat{f}_i)$ is the "Poisson-bracket-commutator" relationship: The Poisson-bracket (2.1.6) gives an algebraic structure on the space of classical observables $\mathcal{O}$, whereas commutators provide a similar algebraic structure on quantum observables $\hat{\mathcal{O}}$. One seeks a correspondence map $\hat{}: \mathcal{O} \mapsto \hat{\mathcal{O}}$, taking classical observables to quantum observables, such that for any pair of classical observables f, g we have

$$[\hat{f}, \hat{g}] = i\,\widehat{\{f, g\}} \tag{2.2.1}$$

where we choose units where $\hbar = 1$. In fact, it is well known (see, e.g., Chernoff (1981) or Gotay (1980)) that—for theories in which the

quantum theory $(\mathcal{F}, V_\alpha)$.

The key theorem alluded to above is the following:

Theorem 2.2.1 (Stone-von Neumann theorem): Let $(\mathcal{M}, \Omega)$ be a finite dimensional symplectic vector space. Let $(\mathcal{F}, \hat{W}(y))$ and $(\mathcal{F}', \hat{W}'(y))$ be strongly continuous, irreducible, unitary representations of the Weyl relations (2.2.6)-(2.2.7). Then $(\mathcal{F}, \hat{W}(y))$ and $(\mathcal{F}', \hat{W}'(y))$ are unitarily equivalent.

A proof of the Stone-von Neumann theorem can be found, e.g., in Simon (1972) (see theorem 7.5 of that reference).

In ordinary Schrodinger quantum mechanics, the Hilbert space is taken to be $\mathcal{F} = L^2(\mathbb{R}^3)$, the position operator, $\hat{q}_\mu$, is represented by multiplication by q_μ, and the momentum operator, $\hat{p}_\mu$, is represented by $-i\, \partial/\partial q_\mu$. The exponentiation of these operators yields an irreducible representation of the Weyl relations. Thus, the Stone-von Neumann theorem provides a powerful justification for the standard choices of Hilbert space and position and momentum operators used in Schrodinger quantum mechanics: Any other choice yielding an (irreducible) representation of the Weyl relations would yield a physically equivalent theory insofar as measurements of position and momentum are concerned. Note, however, that the Stone-von Neumann theorem does not give guidance as to how to represent observables besides position and momentum. As already stated above, a natural prescription (preserving the Poisson-bracket-commutator relationship (2.2.1)) exists for quantum observables corresponding to classical observables which are at most linear in the momentum. However, "factor ordering" ambiguities arise for more general observables and, as mentioned above, no choice of factor ordering can be made so as to satisfy the Poisson-bracket-commutator relationship. Thus, although considerable effort has been devoted to analysis of this issue (see, e.g., Woodhouse (1980)), the question of how to represent general observables in quantum theory remains largely unanswered. Fortunately, these difficulties will not concern us here, since the only observable aside from position and momentum which we shall consider at this stage is a Hamiltonian operator of the form (2.1.11), for which no factor ordering ambiguities arise.

Finally, we emphasize that in the Stone-von Neumann theorem the hypothesis that $\mathcal{M}$ is finite dimensional is essential. As we shall see explicitly in section 4.4, for an infinite dimensional phase space (i.e., for a field system), infinitely many unitarily inequivalent irreducible representations of the Weyl relations exist. For fields in flat spacetime, the additional criterion of Poincaré invariance can be imposed to obtain a preferred representation. However, in a general curved spacetime, no such additional criterion exists. Nevertheless, as we shall see, the algebraic framework will allow us to formulate quantum field theory in curved spacetime without the need of selecting a preferred representation.

2.3 Harmonic Oscillators

We turn our attention, now, to the quantum theory of a classical, linear, dynamical system of finitely many degrees of freedom, i.e., a collection of coupled harmonic oscillators (with possibly time-dependent masses and spring constants). Harmonic oscillators, of course, are among the most widely studied systems in quantum mechanics, and the reader undoubtedly will be very familiar with the standard construction of the quantum theory of time-independent harmonic oscillators. The first task undertaken in this section will be to suitably rewrite this standard quantum theory of time-independent oscillators in a manner that can be generalized to the case of infinitely many degrees of freedom. This reformulation then will be generalized to the case of a general, linear, dynamical system, with time-dependent Hamiltonian, thereby laying the foundation for our construction of quantum field theory in a general curved spacetime. This final reformulation, of course, is equivalent to the standard Schrodinger quantum mechanics of time-dependent oscillators, but it is likely to look quite unfamiliar to the reader.

Consider, first, the standard, "textbook" single, time-independent, harmonic oscillator with Lagrangian

$$\mathcal{L} = \frac{1}{2}\dot{q}^2 - \frac{1}{2}\omega^2 q^2 \tag{2.3.1}$$

and corresponding classical Hamiltonian given by

$$H = \frac{1}{2}p^2 + \frac{1}{2}\omega^2 q^2 \tag{2.3.2}$$

with $\omega \neq 0$. In accordance with the remarks following theorem 2.2.1 above, we can construct the quantum theory by taking the Hilbert space, $\mathcal{F}$, to be $L^2(\mathbb{R})$ and representing q and p by operators on $L^2(\mathbb{R})$

in the standard way. We define the Hamiltonian operator by eq. (2.3.2), with q and p replaced by their corresponding quantum operators. No factor ordering ambiguities occur here, and the operator version of eq. (2.3.2) uniquely defines a self-adjoint operator $\hat{H}$. Since we shall not consider other observables besides q, p, and H here, this completes the specification of the quantum theory of the classical oscillator (2.3.1).

Nevertheless, this quantum theory of a harmonic oscillator can be re-written in a very different looking manner, which will be of considerable importance for quantum field theory. The annihilation and creation operators (usually called lowering and raising operators in the context of harmonic oscillators) play a central role in this new description, and I shall now briefly review their construction within the context of the standard quantum theory. In addition, I shall give the Heisenberg form of the evolution equations, which also will be of considerable importance for field theory.

We define the "annihilation operator", a, by

$$a = \sqrt{\frac{\omega}{2}}\, q + i\sqrt{\frac{1}{2\omega}}\, p \tag{2.3.3}$$

where, here and throughout the following, I shall omit writing "^" above quantum observables except in some cases where confusion with classical observables might arise. It then follows directly that we have

$$[a, a^\dagger] = I \tag{2.3.4}$$

$$H = \omega\,(a^\dagger a + \frac{1}{2} I) \tag{2.3.5}$$

and

$$[H, a] = -\omega\, a \tag{2.3.6}$$

From eq. (2.3.6), we see immediately that in the Heisenberg representation, the annihilation operator satisfies

$$\frac{da_H}{dt} = i[H, a_H] = -\, i\omega a_H \tag{2.3.7}$$

where the subscript "H" denotes Heisenberg representation, i.e.,

$$a_H(t) \equiv U_t^{-1}\, a\, U_t \tag{2.3.8}$$

where U_t is the time evolution operator, i.e., $U_t = \exp(-iHt)$. Hence, we obtain

$$a_H(t) = \exp(-i\omega t)\, a \tag{2.3.9}$$

and

$$q_H(t) = \sqrt{\frac{1}{2\omega}}\,(a_H + a_H{}^\dagger)$$

$$= \sqrt{\frac{1}{2\omega}}\,(\exp(-i\omega t)\, a + \exp(i\omega t)\, a^\dagger) \tag{2.3.10}$$

Thus, the annihilation operator, a, in the Schrodinger picture is precisely the "positive frequency part" of the Heisenberg position operator, q_H. The Heisenberg momentum operator, p_H, is given by

$$p_H = \frac{dq_H}{dt} \tag{2.3.11}$$

It is worth noting that eqs. (2.3.10) and (2.3.11) uniquely determine the Hamiltonian operator up to a multiple of the identity operator. Namely, if two different Hamiltonians, H and H', give rise to the same formulas for q_H and p_H, then by the Heisenberg equation of motion, H–H' must commute with q and p, and hence must commute with all Weyl operators W(y). Since ($\mathfrak{F}$, W(y)) is irreducible it then follows from Schur's lemma that H–H' is a multiple of the identity operator. Thus, eqs. (2.3.10) and (2.3.11) are equivalent to a specification of H up to a multiple of the identity operator. This means of specifying H is particularly convenient in field theory, where, as we shall see, analogs of eqs. (2.3.10) and (2.3.11) can be obtained straightforwardly, but an "infinite subtraction" would be needed to define H directly.

As is familiar to any student of quantum mechanics, the ground state, Ψ_0, of the harmonic oscillator satisfies

$$a\,\Psi_0 = 0 \tag{2.3.12}$$

The nth excited state, Ψ_n, is given by

$$\Psi_n = \sqrt{\frac{1}{n!}}\,(a^\dagger)^n\,\Psi_0 \tag{2.3.13}$$

and it satisfies

$$H\Psi_n = (n + \tfrac{1}{2})\,\omega\,\Psi_n \tag{2.3.14}$$

Before rewriting the quantum theory in a new form, we generalize the above considerations to a collection of n decoupled, time-independent harmonic oscillators of frequencies $\omega_1,\ldots,\omega_n$. According to the standard rules of quantum theory, the Hilbert space, $\mathscr{F}$, of the combined system is taken to be the tensor product (see Appendix A.2) of the Hilbert spaces, $\mathscr{F}_1,\ldots,\mathscr{F}_n$, of the individual oscillator systems,

$$\mathscr{F} = \mathscr{F}_1 \otimes \ldots \otimes \mathscr{F}_n \cong L^2(\mathscr{Q}) \qquad (2.3.15)$$

where $\mathscr{Q}$ is the classical configuration space of the collection of oscillators. The operators q_i, p_k of the individual oscillators extend to $\mathscr{F}$ in an obvious way and yield an irreducible representation of the Weyl relations (2.2.6), (2.2.7). Annihilation and creation operators a_i, $a_i{}^\dagger$ can be introduced for each oscillator as above, and analogs of all the results for the single harmonic oscillator carry over straightforwardly. In particular, the ground state, Ψ_0, satisfies $a_i\Psi_0 = 0$ for all i, and the states obtained by applying arbitrary products of creation operators to Ψ_0 span $\mathscr{F}$.

We now give an alternate construction of the quantum theory of n decoupled, time independent harmonic oscillators. This construction will play a key role in our formulation of quantum field theory, since the tensor product construction (2.3.15) will not generalize suitably to the case of infinitely many degrees of freedom. In addition, this reformulation will be of importance for the development of a particle interpretation of quantum field theory.

We shall formulate the alternative construction in terms of the symplectic vector space $(\mathscr{S}, \Omega)$ of classical solutions of the equations of motion (see the end of section 2.1 above) rather than in terms of phase space, $\mathscr{M}$. (However, we will continue to pass freely between $\mathscr{S}$ and $\mathscr{M}$ by identifying a solution with its initial data at t=0.) Our first step in the construction is to complexify $\mathscr{S}$, thereby obtaining a 2n-complex dimensional vector space, denoted $\mathscr{S}^{\mathbb{C}}$. We extend Ω to $\mathscr{S}^{\mathbb{C}}$ by (complex) linearity in each variable. Now, consider the map $(\ ,\ ): \mathscr{S}^{\mathbb{C}} \times \mathscr{S}^{\mathbb{C}} \rightarrow \mathbb{C}$ defined by

$$(y_1, y_2) = -i\Omega(\overline{y}_1, y_2) \qquad (2.3.16)$$

where the bar denotes complex conjugation. Then $(\ ,\ )$ satisfies all the properties of an inner product (see Appendix A.1) on $\mathscr{S}^{\mathbb{C}}$, except that it fails to be positive definite. However, let $\mathcal{H}$ denote the n-complex dimensional subspace of $\mathcal{S}^{\mathbb{C}}$ spanned by the (complex) solutions of the classical equations of motion which oscillate with purely positive frequency:

$$q_i(t) = \alpha_i \exp(-i\omega_i t) \tag{2.3.17}$$

Then, it is easily verified that (,) is positive definite on $\mathcal{H}$, thus making $\mathcal{H}$ into an n-dimensional complex Hilbert space. Additional key properties of $\mathcal{H}$ will be pointed out below.

Let $\mathcal{F}_s(\mathcal{H})$ denote the symmetric Fock space associated with $\mathcal{H}$, constructed in the manner explained in Appendix A.2. Let $\xi_i \in \mathcal{H}$ denote the (complex) positive frequency solution of the classical equations of motion in which only the ith harmonic oscillator is excited, with ξ_i normalized so that $\|\xi_i\|^2 = (\xi_i, \xi_i) = 1$. Then $\{\xi_i\}$ comprises an orthonormal basis of $\mathcal{H}$. Let a_i denote the annihilation operator on $\mathcal{F}_s(\mathcal{H})$ associated with $\bar{\xi}_i$ (see Appendix A.3).

The new formulation of the quantum theory of our collection of n harmonic oscillators is defined as follows: We choose the Hilbert space to be $\mathcal{F}' = \mathcal{F}_s(\mathcal{H})$. We define the Heisenberg position and momentum operators on $\mathcal{F}'$ by

$$q'_{iH}(t) = \xi_i(t)a_i + \bar{\xi}_i(t)a_i^\dagger \tag{2.3.18}$$

$$p'_{iH}(t) = dq'_{iH}/dt \tag{2.3.19}$$

As pointed out above, eqs. (2.3.18) and (2.3.19) are equivalent to specifying the Schrodinger position and momentum operators together with a specification of the Hamiltonian, H', up to a (physically irrelevant) multiple of the identity operator. Thus, our construction of the Hilbert space $\mathcal{F}'$ together with eqs. (2.3.18)-(2.3.19) define a quantum theory of the system of harmonic oscillators.

Although our construction of ($\mathcal{F}'$; q'_i, p'_j, H') is rather different in appearance from the preceding construction of ($\mathcal{F}$; q_i, p_j, H), it can be readily verified that the two constructions are unitarily equivalent. First, note that from the commutation relations for annihilation and creation operators on Fock space, it may be verified directly that the Schrodinger position and momentum operators determined by this construction satisfy the canonical commutation relations (2.2.2)-(2.2.3). The unitary equivalence of ($\mathcal{F}'$; q'_i, p_j') with ($\mathcal{F}$; q_i, p_j) then follows immediately from the Stone-von Neumann theorem, since it is easily seen that both constructions yield an irreducible representation of the Weyl relations. Since, by eqs. (2.3.10)-(2.3.11) and (2.3.18)-(2.3.19), this unitary correspondence also maps the Heisenberg position and momentum operators into each other, it follows that it must also carry H into H' (up to a multiple of the identity). Indeed, it is easily seen that the unitary correspondence, $U:\mathcal{F} \to \mathcal{F}'$, of the two constructions is given explicitly by $U\Psi_0 = |0\rangle$, $U\Psi_{1i} = a_i^\dagger |0\rangle$, etc., where $|0\rangle$ denotes the "vacuum vector" of the Fock space.

We now rewrite eqs. (2.3.18) and (2.3.19) in a basis-independent manner as follows. Let $\overline{\mathcal{H}}$ denote the complex conjugate space to $\mathcal{H}$ (see Appendix A.2). Then $\overline{\mathcal{H}}$ may be identified with the subspace of negative frequency (complex) solutions to the classical harmonic oscillator equations of motion. Since $\mathcal{H}$ and $\overline{\mathcal{H}}$ are n-complex-dimensional subspaces of $\mathcal{S}^{\mathbb{C}}$ with no nonzero vectors in common, it follows that every $z \in \mathcal{S}^{\mathbb{C}}$ can be expressed uniquely as $z = z^+ + z^-$, with $z^+ \in \mathcal{H}$, $z^- \in \overline{\mathcal{H}}$. This fact allows us to "project" any vector in $\mathcal{S}^{\mathbb{C}}$ onto $\mathcal{H}$. In particular, it follows that there exists a real-linear, one-to-one, and onto "projection map" $K: \mathcal{S} \to \mathcal{H}$, defined by the operation of extracting the "positive frequency part" of any real classical solution. The (basis independent) information corresponding to that contained in eq. (2.3.3) for a single oscillator now can be stated as follows: For each $y \in \mathcal{S}$ the (Schrodinger picture) operator representing the classical observable $\Omega(y,\cdot)$ (see section 2.1) is given by

$$\hat{\Omega}(y,\cdot) = ia(\overline{Ky}) - ia^\dagger(Ky) \qquad (2.3.20)$$

where $a(\overline{Ky})$ denotes the annihilation operator associated with $Ky \in \mathcal{H}$ (see Appendix A.3). Equations (2.3.18) and (2.3.19) then can be re-expressed as the statement that the Heisenberg operator at time t corresponding to the Schrodinger operator (2.3.20) is

$$\hat{\Omega}_H(y,\cdot) = ia(\overline{Ky_t}) - ia^\dagger(Ky_t) \qquad (2.3.21)$$

where y_t denotes the solution whose initial data at time t is the same as the initial data of y at time t = 0 (i.e., y_t is the "time translate" of y by t).

We turn our attention, now, to the quantum theory of a general, linear, dynamical system, described classically by the Hamiltonian (2.1.11). When H is time-independent, we can decouple the degrees of freedom by a normal mode decomposition and apply our prescription above, so our concern here is in the generalization of the above constructions to the time-dependent case. There is no difficulty whatsoever in straightforwardly applying the standard, Schrodinger quantum mechanical construction of ($\mathcal{F}$; q_i, p_j, H) to the case of a general linear dynamical system with time-dependent Hamiltonian. However, this will not be useful for our construction of quantum field theory, since, as we remarked above, the tensor product construction (2.3.15) does not generalize suitably to the case of infinitely many degrees of freedom. On the other hand, it is not immediately obvious how to generalize our construction of ($\mathcal{F}'$; q'_i, p'_j, H'), since that construction made essential use of the time translation symmetry present in the classical theory: If H fails to be time-independent, solutions which oscillate with purely positive frequency (see eq. (2.3.17)) will not exist, so it is not at all clear how to define an analog of the subspace, $\mathcal{H}$, of $\mathcal{S}^{\mathbb{C}}$. It is essential that we reformulate our alternate construction without making use of time translation symmetry if we wish to treat a quantum field in a general, nonstationary, curved spacetime. We now shall proceed to do so.

Let ($\mathcal{S}$, Ω) be the symplectic vector space of solutions to the equations of motion of a general, linear classical system with time-dependent Hamiltonian, H, of the form eq. (2.1.11). As in the case of a time-independent Hamiltonian, we complexify $\mathcal{S}$ and define a (non-positive-definite) inner product on $\mathcal{S}^{\mathbb{C}}$ by eq. (2.3.16). However, as already noted above, in the case of a time-dependent Hamiltonian, solutions do not oscillate harmonically, and there is no natural, unique analog of the subspace of positive frequency solutions used above. Nevertheless, we can proceed by choosing *any* subspace, $\mathcal{H}$, of $\mathcal{S}^{\mathbb{C}}$ which satisfies the following properties:

(i) The "inner product", eq. (2.3.16), is positive definite on $\mathcal{H}$, thus making $\mathcal{H}$ into a Hilbert space over $\mathbb{C}$.

(ii) $\mathcal{S}^{\mathbb{C}}$ is equal to the span of $\mathcal{H}$ and its complex conjugate space, $\overline{\mathcal{H}}$.

(iii) For all $z^+ \in \mathcal{H}$ and $z^- \in \overline{\mathcal{H}}$, we have $(z^+, z^-) = 0$.

It is not difficult to check that there are many choices of subspace, $\mathcal{H}$, which satisfy these three properties.

These properties imply that every $z \in \mathcal{S}^{\mathbb{C}}$ can be uniquely expressed as $z = z^+ + z^-$, with $z^+ \in \mathcal{H}$, $z^- \in \overline{\mathcal{H}}$. Hence, given a choice of $\mathcal{H}$ satisfying these properties, we again obtain a real-linear, one-to-one, onto map $K: \mathcal{S} \to \mathcal{H}$, defined for all $y \in \mathcal{S}$ by $Ky = y^+$. We note that for all $y_1, y_2 \in \mathcal{S}$, we have

$$
\begin{aligned}
\mathrm{Im}(Ky_1, Ky_2)_{\mathcal{H}} &= -\mathrm{Re}\{\Omega(\overline{Ky}_1, Ky_2)\} \\
&= -\frac{1}{2}\,\Omega(\overline{Ky}_1, Ky_2) - \frac{1}{2}\,\Omega(Ky_1, \overline{Ky}_2) \\
&= -\frac{1}{2}\,\Omega(y_1, y_2)
\end{aligned}
\tag{2.3.22}
$$

where in the last step we used condition (iii) together with the fact that for all $y \in \mathcal{S}$, we have $y = Ky + \overline{Ky}$.

To define the quantum theory, we choose the Hilbert space to be $\mathcal{F}_s(\mathcal{H})$. In analogy with eq. (2.3.21), for each $y \in \mathcal{S}$ we define the quantum Heisenberg observable at time t corresponding to $\Omega(y, \cdot)$ to be

$$
\hat{\Omega}_H(y, \cdot) = ia(\overline{Ky}_t) - ia^\dagger (Ky_t)
\tag{2.3.23}
$$

This construction defines a quantum theory of a general linear dynamical system. However, the construction has the potentially unsatisfactory feature that it appears to depend on our choice of $\mathcal{H}$, and—as already noted above—in the absence of time translation symmetry, no natural, "preferred" choice of $\mathcal{H}$ is available. Fortunately, this turns out not to be a difficulty for the case considered here of a system with finitely many degrees of freedom: By the Stone-von Neumann theorem together with the dynamical behavior (2.3.23) of the Heisenberg operators, it follows that for any choice of $\mathcal{H}$ satisfying (i)-(iii) above—$(\mathcal{F}_s(\mathcal{H}); \hat{\Omega}_H(y, \cdot))$ is unitarily equivalent to the standard tensor product construction of the quantum theory. Thus, in particular, for any two subspaces, $\mathcal{H}$ and $\mathcal{H}'$, of $\mathcal{S}^{\mathbb{C}}$, satisfying (i)-(iii), the theories $(\mathcal{F}_s(\mathcal{H}); \hat{\Omega}_H(y, \cdot))$ and $(\mathcal{F}_s(\mathcal{H}'); \hat{\Omega}'_H(y, \cdot))$ are unitarily equivalent. Indeed, a direct proof of this unitary equivalence will follow from the analysis we shall give in section 4.4 below.

Thus, we obtain a fully satisfactory alternate construction of the quantum theory of a linear dynamical system (with finitely many degrees of freedom) possessing a time-dependent Hamiltonian. However, it is worth noting here that—as we shall see explicitly in section 4.4—the situation changes dramatically when $\mathcal{S}$ is infinite dimensional, since in that case different choices of $\mathcal{H}$ will, in general, yield unitarily inequivalent theories. This poses a potential serious difficulty for the formulation of quantum field theory in curved spacetime. As we shall see in section 4.5 this difficulty can be resolved by formulating the theory via the algebraic approach.

It is worth pointing out that even in the case of n decoupled, time-independent oscillators, we also are free to make a choice, $\mathcal{H}'$, of subspace of $\mathcal{S}^{\mathbb{C}}$ (satisfying (i)-(iii)) different from the standard choice, $\mathcal{H}$, of positive frequency solutions. The Hamiltonian on $\mathcal{F}' = \mathcal{F}_s(\mathcal{H}')$ then will no longer have the simple form (2.3.5) in terms of annihilation and creation operators on $\mathcal{F}_s(\mathcal{H}')$—it will also have "aa" and "$a^\dagger a^\dagger$" terms—and the "vacuum state" $|0\rangle' \in \mathcal{F}_s(\mathcal{H}')$ will not be an eigenstate of the Hamiltonian. Nevertheless, $|0\rangle'$ will possess all of the "kinematical properties" of a ground state. In the literature on atomic physics and optics, a state of the form $|0\rangle'$ usually is referred to as a *squeezed vacuum state*. As a by-product of the analysis given in section 4.4 below (see eq. (4.4.23)), we will obtain an explicit expression for $|0\rangle'$ as a state in the standard Hilbert space $\mathcal{F}_s(\mathcal{H})$.

Finally, we point out a useful way of characterizing the freedom available in the choice of $\mathcal{H}$. First, we note that a specification of a complex subspace, $\mathcal{H}$, of $\mathcal{S}^{\mathbb{C}}$ satisfying (i)-(iii) allows us to obtain a real inner product $\mu: \mathcal{S} \times \mathcal{S} \to \mathbb{R}$ on the original ("uncomplexified") vector space of real solutions as follows: For $y_1, y_2 \in \mathcal{S}$, we define

$$\mu(y_1, y_2) = \mathrm{Re}(Ky_1, Ky_2)_{\mathcal{H}} = \mathrm{Im}\,\Omega(\overline{Ky}_1, Ky_2) \tag{2.3.24}$$

where $K: \mathcal{S} \to \mathcal{H}$ denotes the "projection map" onto $\mathcal{H}$. Note that by eq. (2.3.22), we have for all $y_1, y_2 \in \mathcal{S}$,

$$(Ky_1, Ky_2)_{\mathcal{H}} = \mu(y_1, y_2) - \frac{i}{2}\Omega(y_1, y_2) \tag{2.3.25}$$

Now, by the Schwarz inequality, for all $z_1, z_2 \in \mathcal{H}$ we have

$$\|z_1\|^2 \|z_2\|^2 \geq |(z_1,z_2)|^2 \geq |\mathrm{Im}(z_1,z_2)|^2 \qquad (2.3.26)$$

Hence, writing $z_1 = Ky_1$, $z_2 = Ky_2$, we find that μ must satisfy

$$\mu(y_1,y_1)\,\mu(y_2,y_2) \geq \frac{1}{4}\,[\Omega(y_1,y_2)]^2 \qquad (2.3.27)$$

for all $y_1, y_2 \in \mathcal{S}$. Indeed, since K is one-to-one and onto and since the Schwarz inequality (2.3.26) on $\mathcal{H}$ always can be "saturated", we obtain the following stronger version of eq. (2.3.27): For each $y_1 \in \mathcal{S}$ we have

$$\mu(y_1,y_1) = \frac{1}{4} \max_{y_2 \neq 0} \frac{[\Omega(y_1,y_2)]^2}{\mu(y_2,y_2)} \qquad (2.3.28)$$

Note that μ appears on both sides of eq. (2.3.28), and this equation provides a condition on μ which does not determine μ uniquely.

Conversely, suppose we are given a positive-definite, symmetric, bilinear map $\mu: \mathcal{S} \times \mathcal{S} \to \mathbb{R}$ satisfying eq. (2.3.28). Then, for each $y_1 \in \mathcal{S}$, there exists a unique $y_2 \in \mathcal{S}$ such that

$$\frac{1}{2}\,\Omega(y_1,y_2) = \mu(y_1,y_1) \qquad (2.3.29)$$

and

$$\mu(y_2,y_2) = \mu(y_1,y_1) \qquad (2.3.30)$$

(Uniqueness of y_2 follows from the fact that if $y_2' \neq y_2$ also satisfied eqs. (2.3.29) and (2.3.30), then linear combinations of y_2 and y_2' would violate eq. (2.3.28).) The association

$$y_1 \to \frac{1}{2}\,(y_1 + iy_2) \qquad (2.3.31)$$

then maps $\mathcal{S}$ onto an n-complex-dimensional subspace, $\mathcal{H}$, of $\mathcal{S}^{\mathbb{C}}$. It is not difficult to show that $\mathcal{H}$ satisfies properties (i)-(iii) (see the discussion following eq. (3.2.16) below). Furthermore, this choice of $\mathcal{H}$ reproduces the given μ via eq. (2.3.24). Thus, we see that a specification of $\mathcal{H}$ satisfying properties (i)-(iii) is completely equivalent to a specification of real inner product, μ, on $\mathcal{S}$ satisfying eq. (2.3.28). As we shall see, in the field theory case there are significant technical advantages to specifying the quantum theory via a choice of μ rather than via a choice of $\mathcal{H}$.

3 Quantum Fields in Flat Spacetime

This chapter constructs the quantum theory of a free (i.e., linear) Klein-Gordon field in Minkowski spacetime. Section 3.1 presents the construction in a manner that corresponds closely to what is found in standard textbook treatments. This approach relies heavily upon a "plane wave expansion", in which the field assumes the explicit form of an infinite system of decoupled, time-independent, harmonic oscillators. Section 3.2 reformulates this construction without using a plane wave expansion, thereby enabling us to distinguish clearly between the essential input in the construction (namely, the choice of the subspace of "positive frequency solutions" to the Klein-Gordon equation) and inessential input (namely, the choice of a particular basis for these positive frequency solutions). A generalization of this construction which allows one to make more general choices of a subspace of (complex) solutions to the Klein-Gordon equation then will be presented. Since no analog of either a plane wave basis or a choice of "positive frequency subspace" is available in a general curved spacetime, this reformulation is of crucial importance for the construction of quantum field theory in curved spacetime given in the next chapter. Indeed, the reformulation given in section 3.2 comprises most of the "hard work" related to the construction of quantum field theory in curved spacetime. The chapter concludes in section 3.3 with an analysis showing how a particle interpretation can be given to the states of the Klein-Gordon field in Minkowski spacetime. This analysis generalizes straightforwardly to stationary, curved spacetimes (see section 4.3), but in a general, nonstationary spacetime, the states of the quantum field typically will not possess a physically meaningful particle interpretation, except in an approximate or asymptotic sense.

3.1 Basic Construction

In this section, we shall construct the quantum theory of a real, linear (i.e., non-self-interacting) Klein-Gordon field in

Minkowski spacetime. We shall do so by "putting the field in a box", noting that its normal modes comprise an infinite system of decoupled harmonic oscillators, and then applying an appropriate version of the construction of the previous chapter.

The action of the Klein-Gordon field in Minkowski spacetime is

$$S = -\frac{1}{2} \int (\partial_a \phi \, \partial^a \phi + m^2 \phi^2) \, d^4x \tag{3.1.1}$$

The classical equations of motion arising from S are

$$\partial^a \partial_a \phi - m^2 \phi = 0 \tag{3.1.2}$$

By introducing a global inertial coordinate system, we can re-write eq. (3.1.1) as

$$S = \int \mathcal{L} \, dt \tag{3.1.3}$$

with

$$\mathcal{L} = \frac{1}{2} \int [\dot{\phi}^2 - (\vec{\nabla} \phi)^2 - m^2 \phi^2] \, d^3x \tag{3.1.4}$$

where the dot denotes a time derivative.

In order to simplify the technical details of the initial discussion, we now (provisionally) replace the infinite Euclidean space, $\mathbb{R}^3$, by a flat 3-torus, T^3, of side L, i.e., we "put the field in a box with periodic boundary conditions." A reformulation of this construction without using this artifice will be given in the next section. We then expand ϕ in a Fourier series,

$$\phi(t,\mathbf{x}) = L^{-3/2} \sum_{\mathbf{k}} \phi_{\mathbf{k}}(t) \exp(i\mathbf{k} \cdot \mathbf{x}) \tag{3.1.5}$$

where the vector **k** is of the form

$$\mathbf{k} = \frac{2\pi}{L} (n_1, n_2, n_3) \tag{3.1.6}$$

with n_1, n_2, n_3 integers, and where, in this section, boldface letters will indicate (three-dimensional) spatial vectors. The coefficients $\phi_{\mathbf{k}}(t)$ are given in terms of ϕ by

$$\phi_{\mathbf{k}}(t) = L^{-3/2} \int \phi(t,\mathbf{x}) \exp(-i\mathbf{k} \cdot \mathbf{x}) \, d^3x \tag{3.1.7}$$

The reality of ϕ implies that

$$\overline{\phi}_{\mathbf{k}} = \phi_{-\mathbf{k}} \tag{3.1.8}$$

The Lagrangian, $\mathcal{L}$, then can be expressed as

$$\mathcal{L} = \sum_{\mathbf{k}} \left\{ \frac{1}{2} |\dot{\phi}_{\mathbf{k}}|^2 - \frac{1}{2} \omega_{\mathbf{k}}^2 |\phi_{\mathbf{k}}|^2 \right\} \tag{3.1.9}$$

where

$$\omega_{\mathbf{k}}^2 = \mathbf{k}^2 + m^2 \tag{3.1.10}$$

Comparing eq. (3.1.9) with eq. (2.3.1), we see that the Klein-Gordon field may be viewed as an infinite collection of decoupled, time-independent, harmonic oscillators. More precisely, $\sqrt{2}$ Re $(\phi_{\mathbf{k}})$ and $\sqrt{2}$ Im $(\phi_{\mathbf{k}})$ are the "position observables" for a countably infinite set of independent, real harmonic oscillators. (Here the factor of $\sqrt{2}$ arises from the fact that the oscillator labeled by $\mathbf{k}$ is the same oscillator as the one labeled by $-\mathbf{k}$ on account of eq. (3.1.8), so the sum over all $\mathbf{k}$ in eq. (3.1.9) double-counts the oscillators.) Thus, in order to construct the quantum field theory of the free Klein-Gordon scalar field, we simply need to generalize the construction of the previous chapter for finitely many decoupled, time independent oscillators to the case of a countably infinite number of oscillators.

In the previous chapter, we gave two equivalent prescriptions for constructing the Hilbert space for the quantum theory of finitely many oscillators: (1) Take the tensor product space of the Hilbert spaces for the individual oscillators. (2) Construct a Fock space based upon a Hilbert space of (complex) classical solutions. In the present case of infinitely many oscillators, this first prescription is no longer available. Although the tensor product of infinitely many Hilbert spaces can be defined (von Neumann 1938), it is "too large" to be suitable for use in a (normal type of) quantum theory in that it is not separable and the representation it provides of the canonical commutation relations is reducible.

On the other hand, the second prescription generalizes straightforwardly to the present case. We define $\mathcal{H}$ to be the Hilbert space of positive frequency solutions to the Klein-Gordon equation (3.1.2), with inner product defined by eq. (2.3.16). In the next section, we shall state more precisely and generally how $\mathcal{H}$ is constructed, but in the present context (where "space" is T^3), it suffices to say that an orthonormal basis of $\mathcal{H}$ is provided by "plane-wave" solutions of the form

$$\psi_{\mathbf{k}} = \frac{\exp(i\mathbf{k}\cdot\mathbf{x} - i\omega_{\mathbf{k}}t)}{(2\omega_{\mathbf{k}})^{1/2}L^{3/2}} \tag{3.1.11}$$

with $\mathbf{k}$ given by eq. (3.1.6) and $\omega_{\mathbf{k}} > 0$ given by eq. (3.1.10).

We choose the Hilbert space, $\mathcal{F}$, to be $\mathcal{F}_s(\mathcal{H})$. (The fact that $\mathcal{H}$ is now infinite dimensional poses no problem for the definition of $\mathcal{F}_s(\mathcal{H})$; see Appendix A.2 for further discussion of the Fock space construction.) Let $b_{\mathbf{k}}$ and $c_{\mathbf{k}}$ denote, respectively, the annihilation operators associated with the normalized positive frequency solutions corresponding to excitations of the oscillators $\sqrt{2}\,\mathrm{Re}\,(\phi_{\mathbf{k}})$ and $\sqrt{2}\,\mathrm{Im}\,(\phi_{\mathbf{k}})$, respectively. (By the reality condition (3.1.8), $b_{\mathbf{k}}$ and $c_{\mathbf{k}}$ are not independent of $b_{-\mathbf{k}}$ and $c_{-\mathbf{k}}$; indeed, we have $b_{-\mathbf{k}} = b_{\mathbf{k}}$, $c_{-\mathbf{k}} = -c_{\mathbf{k}}$.) The "position observables" of the theory are then represented in the Schrodinger picture by the operators

$$\sqrt{2}\,\mathrm{Re}\,\phi_{\mathbf{k}} = \frac{1}{\sqrt{2\omega_{\mathbf{k}}}}\,(b_{\mathbf{k}} + b_{\mathbf{k}}^{\dagger}) \tag{3.1.12}$$

$$\sqrt{2}\,\mathrm{Im}\,\phi_{\mathbf{k}} = \frac{1}{\sqrt{2\omega_{\mathbf{k}}}}\,(c_{\mathbf{k}} + c_{\mathbf{k}}^{\dagger}) \tag{3.1.13}$$

Analogs of eqs. (2.3.18) and (2.3.19) then hold for the Heisenberg position and momentum operators. This specifies the key observables—namely, "position", "momentum", and the Hamiltonian up to a multiple of the identity operator—of the quantum field theory. It should be emphasized that there are no mathematical difficulties whatsoever at this stage of the construction.

It is convenient to define the operator $a_{\mathbf{k}}$ by

$$a_{\mathbf{k}} = \frac{1}{\sqrt{2}}\,(b_{\mathbf{k}} + ic_{\mathbf{k}}) \tag{3.1.14}$$

Then $a_{\mathbf{k}}$ and $a_{-\mathbf{k}}$ can be viewed as independent, and we have (for all $\mathbf{k}$ of the form (3.1.6))

$$[a_{\mathbf{k}}, a_{\mathbf{k}'}] = 0,\ [a_{\mathbf{k}}, a_{\mathbf{k}'}^{\dagger}] = \delta_{\mathbf{k}\mathbf{k}'}\,I \tag{3.1.15}$$

It follows directly from eqs. (3.1.12)-(3.1.14) that the Schrodinger picture observable $\hat{\phi}_{\mathbf{k}}$ corresponding to the classical complex observable $\phi_{\mathbf{k}}$ defined by eq. (3.1.7) is given by

$$\hat{\phi}_{\mathbf{k}} = \frac{1}{\sqrt{2\omega_{\mathbf{k}}}}\,(a_{\mathbf{k}} + a_{-\mathbf{k}}{}^{\dagger}) \tag{3.1.16}$$

Equations (3.1.5) and (3.1.16) now yield the following formal expression for a (Schrodinger picture) operator $\hat{\phi}(\mathbf{x})$ representing the value of the field at spatial point $\mathbf{x}$:

$$\begin{aligned}\hat{\phi}(\mathbf{x}) &= L^{-3/2}\sum_{\mathbf{k}} \exp(i\mathbf{k}\cdot\mathbf{x})\,\hat{\phi}_{\mathbf{k}} \\ &= \sum_{\mathbf{k}}\Big\{\frac{\exp(i\mathbf{k}\cdot\mathbf{x})}{(2\omega_{\mathbf{k}})^{1/2}L^{3/2}}\,a_{\mathbf{k}} + \frac{\exp(-i\mathbf{k}\cdot\mathbf{x})}{(2\omega_{\mathbf{k}})^{1/2}L^{3/2}}\,a_{\mathbf{k}}{}^{\dagger}\Big\}\end{aligned} \tag{3.1.17}$$

The corresponding Heisenberg operator $\hat{\phi}(t;\mathbf{x})$ is given by

$$\hat{\phi}(t;\mathbf{x}) = \sum_{\mathbf{k}}\Big\{\frac{\exp(i\mathbf{k}\cdot\mathbf{x} - i\omega_{\mathbf{k}}t)}{(2\omega_{\mathbf{k}})^{1/2}L^{3/2}}\,a_{\mathbf{k}} + \text{h.c.}\Big\} \tag{3.1.18}$$

where "h.c." denotes "hermitian conjugate", i.e., the adjoint of the preceding operator. Note that eq. (3.1.18) has the form

$$\hat{\phi}(t;\mathbf{x}) = \sum_{i}\{\psi_i(t,\mathbf{x})\,a_i + \overline{\psi}_i(t,\mathbf{x})\,a_i{}^{\dagger}\} \tag{3.1.19}$$

where $\{\psi_i\}$ is the orthonormal basis (3.1.11) of $\mathcal{H}$, and a_i is the annihilation operator associated with ψ_i. However, although all of the steps preceding eq. (3.1.17) are mathematically well defined, the infinite sums appearing in eqs. (3.1.17)-(3.1.19) do not converge, and the theory actually does not admit an observable corresponding to the value of the field at spacetime point $(t, \mathbf{x})$. Nevertheless, as we shall see at the end of the next section, a well defined expression can be given for the quantum field as an operator-valued distribution, which formally corresponds to eq. (3.1.19).

3.2 Reformulation

In the previous section, we constructed a quantum theory of a real Klein-Gordon scalar field by using a plane wave basis of solutions to explicitly express the field as a collection of decoupled, time-independent harmonic oscillators. In this section we shall reformulate this construction in a manner that does not rely upon having such a plane wave basis. As already mentioned above, this reformulation is essential for generalizing the construction to curved spacetime, where no analog of such a plane wave basis is available in general. (Indeed, such a reformulation also is needed to make the construction rigorous in flat spacetime when we do not "put the field in a box", since plane waves in $\mathbb{R}^3$ are not normalizable.) We then shall define (distributional) field observables in the theory corresponding to eq. (3.1.19).

To begin, we must define more precisely the classical phase space, $\mathcal{M}$, and solution space, $\mathcal{S}$, of the Klein-Gordon field. From the Lagrangian density (3.1.4), we see that the momentum density, π, of the Klein-Gordon field on the three-dimensional hypersurface, Σ_0, corresponding to global inertial time t=0 is given by

$$\pi = \frac{\delta S}{\delta \dot{\phi}} = \dot{\phi} \tag{3.2.1}$$

Thus, a point in phase space corresponds to a specification of the pair of functions $\phi(x)$ and $\pi(x)$ on Σ_0. However, in order to define $\mathcal{M}$ precisely, we must specify precisely what class of functions are permitted. This is largely at our discretion, limited only by the following considerations: We need a sufficiently "nice" class of functions in order that mathematical structures such as the symplectic form (see below) are well defined. However, the choice of functions cannot be so restrictive that the local degrees of freedom of the system are inhibited. The class of functions comprised by "Schwartz space" (see, e.g., Reed and Simon (1980)) meets these criteria and is most commonly chosen for quantum field theory in Minkowski spacetime. However, the notion of Schwartz space does not generalize to a general, curved spacetime. Hence, we shall define $\mathcal{M}$ to consist of initial data which are smooth and of compact support on Σ_0, i.e.,

$$\mathcal{M} \equiv \{[\phi, \pi] \mid \phi:\Sigma_0 \rightarrow \mathbb{R},\ \pi:\Sigma_0 \rightarrow \mathbb{R};\ \phi,\pi \in C_0^\infty(\Sigma_0)\} \tag{3.2.2}$$

By the well posedness of the initial value formulation for eq. (3.1.2) (see section 4.1 below), it follows that every point of $\mathcal{M}$ uniquely determines a solution to the Klein-Gordon equation. We define $\mathcal{S}$ to be the space of solutions which arise from initial data in $\mathcal{M}$.

The Lagrangian (3.1.4) gives rise to a symplectic structure $\Omega:\mathcal{M}\times\mathcal{M}\to\mathbb{R}$ via

$$\Omega\{[\phi_1, \pi_1], [\phi_2, \pi_2]\} = \int_{\Sigma_0} d^3x\, (\pi_1 \phi_2 - \pi_2 \phi_1) \tag{3.2.3}$$

Again, the "symplectic product" is conserved for solutions, thus giving rise to a natural symplectic structure (also denoted Ω) on $\mathcal{S}$.

The fundamental Poisson brackets on $\mathcal{M}$ can be expressed as (see eq. (2.1.10) above)

$$\left\{\Omega([\phi_1, \pi_1],\cdot), \Omega([\phi_2, \pi_2],\cdot)\right\} = -\,\Omega([\phi_1, \pi_1], [\phi_2, \pi_2]) \tag{3.2.4}$$

Note that the significant advantage of eq. (2.1.10) over eqs. (2.1.7)-(2.1.8) now can be seen, since no natural choice of basis of $\mathcal{M}$ exists; in particular, the (real and imaginary parts of) plane wave solutions do not lie in $\mathcal{M}$. Note also that if we choose $[\phi_1, \pi_1] = [0, f_1]$ and $[\phi_2, \pi_2] = [f_2, 0]$, eq. (3.2.4) takes the form

$$\left\{\int f_1\phi, \int f_2\pi\right\} = \int f_1 f_2 \tag{3.2.5}$$

(Here "$\int f_1\phi$" means the function on $\mathcal{M}$ whose value at point $[\phi, \pi]$ is $\int f_1\phi$.) Equation (3.2.5) can be re-expressed more loosely in the familiar form

$$\left\{\phi(x_1), \pi(x_2)\right\} = \delta(x_1 - x_2) \tag{3.2.6}$$

We wish to construct a quantum theory in which the functions $\Omega([\phi,\pi],\cdot)$ on the classical phase space $\mathcal{M}$ are represented (irreducibly) by operators $\hat{\Omega}([\phi,\pi],\cdot)$ satisfying the commutation relations corresponding to (3.2.4), namely,

$$\left[\hat{\Omega}([\phi_1, \pi_1],\cdot), \hat{\Omega}([\phi_2, \pi_2],\cdot)\right] = -i\, \Omega([\phi_1, \pi_1], [\phi_2, \pi_2])\, I \tag{3.2.7}$$

As in the finite dimensional case, we will freely make use of the correspondence between $\mathcal{M}$ and $\mathcal{S}$. Thus, equivalently, we seek to obtain operators $\hat{\Omega}(\psi,\cdot)$ satisfying the analog of eq. (3.2.7) for all $\psi \in \mathcal{S}$. Motivated by the ideas and results of the previous subsection (as well as those for the finite degree of freedom case), we construct the quantum theory as follows: Given any $\psi \in \mathcal{S}$, we decompose it into its parts which oscillate with positive and negative frequency with respect to time:

$$\psi = \psi^+ + \psi^- \tag{3.2.8}$$

(A mathematically precise prescription for this decomposition will be given in section 4.3 in the more general context of a Klein-Gordon field in a curved, stationary spacetime.) Let $\mathcal{S}^{\mathbb{C}+}$ denote the subspace spanned by the positive frequency parts of solutions in $\mathcal{S}$. (Note, however, that since a positive frequency solution can be extended to a holomorphic function in the lower-half complex-t plane, it follows from the "edge of the wedge" theorem (see Streater and Wightman (1964)) that any nonzero element of $\mathcal{S}^{\mathbb{C}+}$ cannot have initial data of compact support, and hence cannot lie in the complexification, $\mathcal{S}^{\mathbb{C}}$, of $\mathcal{S}$. The awkwardness caused by this would have been alleviated if we had chosen $\mathcal{S}$ to be the space of solutions with initial data in Schwartz space.) On $\mathcal{S}^{\mathbb{C}+}$, we define the Klein-Gordon inner product by

$$(\psi^+, \chi^+) = -i\Omega(\overline{\psi}^+, \chi^+) \tag{3.2.9}$$

Here the definition of Ω has been extended to $\mathcal{S}^{\mathbb{C}+}$ using eq. (3.2.3), and it can be shown that this extension is well defined and that the right side of eq. (3.2.9) is positive-definite, provided that, in the case of a massless Klein-Gordon field, the Minkowski spacetime dimension is greater than two. ("Infra-red divergences" occur for the massless Klein-Gordon field in two spacetime dimensions, i.e., in general, ψ^+ will fall off too slowly at infinity to allow Ω to be defined.) We then Cauchy-complete $\mathcal{S}^{\mathbb{C}+}$ in the norm defined by this inner product to obtain a complex Hilbert space $\mathcal{H}$ (see Appendix A.1). This provides a mathematically precise definition of $\mathcal{H}$ without the need to "put the field in a box". Note that the association $\psi \to \psi^+$ yields a (real) linear one-to-one map $K: \mathcal{S} \to \mathcal{H}$ which takes $\mathcal{S}$ into a dense subspace of $\mathcal{H}$.

The quantum theory is specified as follows: We take the Hilbert space to be $\mathcal{F}_s(\mathcal{H})$. For each $\psi \in \mathcal{S}$, we define the operator $\hat{\Omega}(\psi, \cdot)$ on $\mathcal{F}_s(\mathcal{H})$ by

$$\hat{\Omega}(\psi,\cdot) = ia(\overline{K\psi}) - ia^\dagger(K\psi) \tag{3.2.10}$$

(see eq. (2.3.20) above). We take the Heisenberg representative of this observable to be

$$\hat{\Omega}_H(\psi,\cdot) = ia(\overline{K\psi_t}) - ia^\dagger(K\psi_t) \tag{3.2.11}$$

(see eq. (2.3.21)) where ψ_t is the solution whose initial data at time t equals the initial data for ψ at $t = 0$. As before, eq. (3.2.11) determines the Hamiltonian up to a multiple of the identity operator.

To compare this construction with that of the previous section, suppose again that "space" is a 3-torus of side L. Let ψ be the solution whose initial data at $t = 0$ is $[0, L^{-3/2} \cos \mathbf{k}\cdot\mathbf{x}]$. Then by eqs. (3.2.3) and (3.1.7), the value of function $\Omega(\psi,\cdot)$ at a point $[\phi, \pi]$ of $\mathcal{M}$ is just Re $(\phi_\mathbf{k})$. Now, the positive and negative frequency parts of the solution ψ are

$$\psi^+ = i(2\omega_\mathbf{k})^{-1}L^{-3/2} \cos(\mathbf{k}\cdot\mathbf{x}) \exp(-i\omega_\mathbf{k}t) \tag{3.2.12}$$

$$\psi^- = -\, i(2\omega_\mathbf{k})^{-1}L^{-3/2} \cos(\mathbf{k}\cdot\mathbf{x}) \exp(+i\omega_\mathbf{k}t) \tag{3.2.13}$$

Hence, we have

$$a(\overline{K\psi}) = -\frac{i}{2}(2\omega_\mathbf{k})^{-1/2}(a_\mathbf{k} + a_{-\mathbf{k}}) \tag{3.2.14}$$

where $a_\mathbf{k}$ denotes the annihilation operator associated with the normalized solution $(2\omega_\mathbf{k})^{-1/2} L^{-3/2} \exp(i\mathbf{k}\cdot\mathbf{x} - i\omega_\mathbf{k}t)$. Thus, eq. (3.2.10) reduces to

$$\hat{\Omega}(\psi,\cdot) = \frac{1}{2}(2\omega_\mathbf{k})^{-1/2}(a_\mathbf{k} + a_{-\mathbf{k}}) + \text{h.c.} \tag{3.2.15}$$

This agrees precisely with the expression for Re $(\phi_\mathbf{k})$ obtained from eq. (3.1.16). Similar results hold for Im $(\phi_\mathbf{k})$. Thus, our reformulation reproduces the quantum theory constructed in the previous section. However, the necessity of choosing a preferred, plane wave basis now has been completely eliminated, and we also have dispensed with the artifice of "putting the field in a box".

Although we have eliminated the need to choose a plane wave basis, it should be emphasized that there does remain one important choice in our reformulation of the quantum theory: the specification of an appropriate subspace of (complex) solutions which is to serve as the Hilbert space, $\mathcal{H}$. As discussed in section 2.3, in the case of finitely many oscillators, the choice of $\mathcal{H}$ actually is irrelevant in that different choices of $\mathcal{H}$ (satisfying properties (i)-(iii)) give rise to unitarily equivalent theories. Indeed, our choice of $\mathcal{H}$ as the subspace of positive frequency solutions for finitely many, time-independent oscillators is "preferred" only in that it has the convenient feature of making the state $|0\rangle \in \mathcal{F}_s(\mathcal{H})$ a ground state of the Hamiltonian. However, as we shall see explicitly in section 4.4 below, the choice of $\mathcal{H}$ is *not* irrelevant in field theory. If our interest was solely to define quantum field theory in Minkowski spacetime, we could justify our selection of $\mathcal{H}$ as the subspace of positive frequency solutions as being the natural one arising from the time translation invariance of the classical theory, which, in turn, is a consequence of the Poincaré invariance of the theory. However, in a general, nonstationary, curved spacetime, we will not have any criterion available (like Poincaré invariance) to select a unique, preferred choice of $\mathcal{H}$, and we will have to deal with the large class of unitarily inequivalent constructions arising from our freedom to choose $\mathcal{H}$.

Therefore, it is important to understand exactly what freedom is available in the choice of $\mathcal{H}$, i.e., to state precisely what conditions on $\mathcal{H}$ are necessary and sufficient for the above construction of a quantum field theory to be mathematically well defined. In the case of finitely many oscillators, the freedom involved in choosing $\mathcal{H}$ was characterized in a very direct and simple fashion as the freedom of choosing a subspace of $\mathcal{S}^{\mathbb{C}}$ satisfying properties (i)-(iii) of section 2.3. However, when $\mathcal{S}$ is infinite dimensional, there is considerable awkwardness in attempting to characterize $\mathcal{H}$ in this manner. The main reason for this awkwardness is that, unlike the finite-dimensional case, $\mathcal{S}$ must first be suitably "enlarged" (by Cauchy-completion) before $\mathcal{H}$ can be viewed as a subspace of $\mathcal{S}^{\mathbb{C}}$. Indeed, we have already encountered this phenomenon in our construction above: No (nonzero) element of the Hilbert space, $\mathcal{H}$, of the standard Minkowski spacetime quantum field theory actually lies in $\mathcal{S}^{\mathbb{C}}$, since all the solutions in $\mathcal{S}^{\mathbb{C}}$ have initial data of compact support, whereas no positive frequency solution can have this property. However, the required choice of "enlargement" of $\mathcal{S}$ makes use of the inner product of $\mathcal{H}$, and thus is really part of the specification of $\mathcal{H}$.

Consequently, there is considerable potential for "circularity" in attempting to define $\mathcal{H}$ as a subspace of an enlarged $\mathcal{S}^{\mathbb{C}}$, and there is considerable advantage to a construction which, in effect, simultaneously specifies both the enlargement of $\mathcal{S}$ and the choice of $\mathcal{H}$.

A much more satisfactory way of specifying the freedom involved in the choice of $\mathcal{H}$ is obtained by proceeding in the manner described at the end of chapter 2: We specify a real inner product $\mu:\mathcal{S}\times\mathcal{S}\to\mathbb{R}$ satisfying, for all $\psi_1\in\mathcal{S}$,

$$\mu(\psi_1,\psi_1) = \frac{1}{4} \underset{\psi_2\neq 0}{\text{l.u.b.}} \frac{[\Omega(\psi_1,\psi_2)]^2}{\mu(\psi_2,\psi_2)} \tag{3.2.16}$$

Note that eq. (3.2.16) differs from eq. (2.3.28) only in that we have replaced "max" by "least upper bound" (l.u.b.), since, in infinite dimensions, a maximum over all $\psi_2\neq 0$ need not be attained. Given such a μ, we can enlarge $\mathcal{S}$ to a real Hilbert space, $\mathcal{S}_\mu$, with inner product $(\psi_1,\psi_2) = 2\mu(\psi_1,\psi_2)$, by taking its Hilbert space completion (see Appendix A.1) in the norm 2μ. By eq. (3.2.16), the bilinear map, $\Omega:\mathcal{S}\times\mathcal{S}\to\mathbb{R}$, is bounded in the norm 2μ, so we may extend its action to $\mathcal{S}_\mu\times\mathcal{S}_\mu$ by continuity. We then define the operator $J:\mathcal{S}_\mu\to\mathcal{S}_\mu$ by

$$\Omega(\psi_1,\psi_2) = 2\mu(\psi_1,J\psi_2) = (\psi_1,J\psi_2) \tag{3.2.17}$$

(Since Ω is bounded, existence of a bounded linear map, J, on $\mathcal{S}_\mu$ satisfying eq. (3.2.17) follows from the Riesz lemma (see Appendix A.2) applied to a real Hilbert space.) From the antisymmetry of Ω, it follows immediately that $J^\dagger = -J$. However, eq. (3.2.16) is equivalent to the statement that J is norm preserving in the inner product 2μ, i.e., $J^\dagger J = I$. Hence, in particular, we have $J^2 = -I$, and, consequently, J endows $\mathcal{S}_\mu$ with a complex structure. Thus, we have shown that a specification of an inner product, μ, satisfying eq. (3.2.16) gives rise to a complex structure, J, on $\mathcal{S}_\mu$. Note that, conversely, the specification of a complex structure, J, on $\mathcal{S}$ for which $-\Omega(\psi_1,J\psi_2)$ is a positive definite inner product would give rise, via eq. (3.2.17), to a μ satisfying eq. (3.2.16), so one can view the specification of an inner product, μ, on $\mathcal{S}$ satisfying eq. (3.2.16) as being roughly equivalent to the specification of a complex structure, J, on $\mathcal{S}$ satisfying the condition that $-\Omega(\psi_1,J\psi_2)$ is positive definite. However, this equivalence is only rough because only a small class of complex structures obtained from the collection of μ's which satisfy eq.

(3.2.16) would actually map $\mathcal{S}$ into itself (rather than into $\mathcal{S}_\mu$).

We now complexify $\mathcal{S}_\mu$ and extend the actions of Ω, μ, and J from $\mathcal{S}_\mu$ to $\mathcal{S}_\mu{}^{\mathbb{C}}$ by complex linearity. We define a (complex) inner product on $\mathcal{S}_\mu{}^{\mathbb{C}}$ by

$$(\psi_1,\psi_2) = 2\mu(\overline{\psi}_1,\psi_2) \qquad (3.2.18)$$

for all $\psi_1,\psi_2 \in \mathcal{S}_\mu{}^{\mathbb{C}}$, thus making $\mathcal{S}_\mu{}^{\mathbb{C}}$ into a (complex) Hilbert space. To obtain the desired complex Hilbert subspace $\mathcal{H} \subset \mathcal{S}_\mu{}^{\mathbb{C}}$, we note that $iJ: \mathcal{S}_\mu{}^{\mathbb{C}} \to \mathcal{S}_\mu{}^{\mathbb{C}}$ is self-adjoint. From the spectral theorem, it then follows that $\mathcal{S}_\mu{}^{\mathbb{C}}$ decomposes into two orthogonal eigensubspaces of J (with eigenvalues $\pm i$), which are easily seen to be complex conjugates of each other. We define $\mathcal{H} \subset \mathcal{S}_\mu{}^{\mathbb{C}}$ to be the eigensubspace with $J = +i$. It then follows directly that $\mathcal{H}$ satisfies properties (i)-(iii) of section 2.3, with $\mathcal{S}^{\mathbb{C}}$ replaced by $\mathcal{S}_\mu{}^{\mathbb{C}}$ in property (ii). Thus, this construction generalizes to infinite dimensions the prescription for the finite dimensional case given in the last paragraph of section 2.3.

This construction shows that a choice of $\mathcal{H}$ can be specified in a mathematically precise manner by a choice of μ satisfying eq. (3.2.16). Furthermore, note that the construction allows us to define the map $K: \mathcal{S}_\mu{}^{\mathbb{C}} \to \mathcal{H}$ to be the orthogonal projection (in the inner product (3.2.18)) onto the subspace, $\mathcal{H}$, of $\mathcal{S}_\mu{}^{\mathbb{C}}$. The restriction of K to $\mathcal{S}$ defines a real-linear map $K: \mathcal{S} \to \mathcal{H}$ with dense range in $\mathcal{H}$ such that for all $\psi_1,\psi_2 \in \mathcal{S}$,

$$\begin{aligned}(K\psi_1, K\psi_2)_{\mathcal{H}} &= -i\Omega(\overline{K\psi_1}, K\psi_2)\\ &= \mu(\psi_1,\psi_2) - \frac{i}{2}\Omega(\psi_1,\psi_2)\end{aligned} \qquad (3.2.19)$$

In particular, we have,

$$\mathrm{Im}(K\psi_1, K\psi_2)_{\mathcal{H}} = -\frac{1}{2}\Omega(\psi_1,\psi_2) \qquad (3.2.20)$$

Conversely, a specification of a Hilbert space $\mathcal{H}$ together with a real-linear map $K: \mathcal{S} \to \mathcal{H}$ which satisfies eq. (3.2.20) and whose range is dense in $\mathcal{H}$ gives rise, via eq. (3.2.19), to a μ which satisfies eq. (3.2.16). This shows that the freedom involved in choosing $\mathcal{H}$ is completely characterized by a choice of μ which satisfies eq. (3.2.16).

Actually, it is possible to consider even more general constructions in which μ merely satisfies the inequality (2.3.27) and, correspondingly, only the complexification of the range of K restricted to $\mathcal{S}$ is dense in $\mathcal{H}$ (so that the span of $\mathcal{H}$ and its complex conjugate space, $\overline{\mathcal{H}}$, is "larger" than $\mathcal{S}_\mu{}^{\mathbb{C}}$). (Details of this more general construction can be found in Appendix A of Kay and Wald (1991).) Indeed, in the algebraic formulation (see section 4.5 below), it is natural to consider a far wider class of quantum field theory constructions. However, the class of constructions arising from a μ satisfying eq. (3.2.16) comprises the class most closely analogous to the quantum theory of finitely many oscillators, and—until we discuss the algebraic formulation—we shall restrict attention to this class.

Once we have chosen $\mathcal{H}$ via a choice of μ satisfying eq. (3.2.16), the remainder of specification of the quantum field theory construction is straightforward: We take the Hilbert space of the quantum field theory to be $\mathcal{F}_s(\mathcal{H})$. The fundamental observables $\hat{\Omega}(\psi,\cdot)$ on $\mathcal{F}_s(\mathcal{H})$ then are defined by eq. (3.2.10), where K is determined by μ as described above.

We conclude this section by showing that—in any of these constructions (including, of course, the original, standard construction)—the operator (3.2.10) actually can be re-interpreted as a spacetime average of the (Heisenberg) operator representing the value of the field. This will enable us to give a rigorous meaning to eq. (3.1.19), with $\hat{\phi}$ viewed as an "operator-valued distribution".

This re-interpretation arises from a relationship between classical solutions to eq. (3.1.2) and the vector space, $\mathcal{T} = C_0^\infty(\mathbb{R}^4)$, of *test functions* on Minkowski spacetime, i.e., $\mathcal{T}$ is comprised by functions on Minkowski spacetime which are smooth and of compact support. Recall that $(\mathcal{S}, \Omega)$ was defined to be the symplectic vector space of solutions to eq. (3.1.2) with initial data, $[\phi, \pi]$, lying in $C_0^\infty(\Sigma_0)$. Let $f \in \mathcal{T}$ and let Af and Rf denote, respectively, the advanced and retarded solutions to the Klein-Gordon equation with source f, so that

$$(\partial^a\partial_a - m^2)(Af) = f, \quad (\partial^a\partial_a - m^2)(Rf) = f \tag{3.2.21}$$

with Af = 0 outside the causal past of the support of f, and Rf = 0 outside the causal future of the support of f. Then

$$Ef \equiv Af - Rf \tag{3.2.22}$$

is a solution to the homogeneous equation (3.1.2). Furthermore the initial data for this solution lies in $C_0^\infty(\Sigma_0)$. Hence, we obtain a linear map $E:\mathfrak{T} \to \mathcal{S}$. The key properties of this map are stated in the following lemma:

Lemma 3.2.1 : The map $E:\mathfrak{T} \to \mathcal{S}$ satisfies the following three properties. (1) E is onto, i.e., every $\psi \in \mathcal{S}$ can be expressed as $\psi = Ef$ for some $f \in \mathfrak{T}$. (2) $Ef = 0$ if and only if $f = (\partial^a\partial_a - m^2)g$ for some $g \in \mathfrak{T}$. (3) For all $\psi \in \mathcal{S}$ and all $f \in \mathfrak{T}$, we have

$$\int \psi f \, d^4x = \Omega(Ef, \psi) \tag{3.2.23}$$

Proof : To prove property (1), let $\psi \in \mathcal{S}$. Let χ be any smooth function on spacetime such that $\chi = 0$ for $t \leqq 0$ and $\chi = 1$ for $t \geqq 1$, and define

$$f = -(\partial^a\partial_a - m^2)(\chi\psi) \tag{3.2.24}$$

Then $f \in \mathfrak{T}$. Furthermore, $Af = (1 - \chi)\psi$ and $Rf = -\chi\psi$, since these quantities solve the inhomogeneous Klein-Gordon equation with the correct boundary conditions. Hence, we obtain $Ef = \psi$, as we desired to show.

To prove property (2), we note that if $f = (\partial^a\partial_a - m^2)g$, then $Af = Rf = g$, so $Ef = 0$. Conversely, if $Ef = 0$, then $Af = Rf$, so both Af and Rf lie in $\mathfrak{T}$. But we have

$$f = (\partial^a\partial_a - m^2)Af \tag{3.2.25}$$

and thus $f = (\partial^a\partial_a - m^2)g$ with $g = Af \in \mathfrak{T}$.

Finally, to prove property (3), let $\psi \in \mathcal{S}$, $f \in \mathfrak{T}$, and choose $t_1, t_2 \in \mathbb{R}$ so that $f = 0$ for all $t \notin [t_1, t_2]$. Then, integrating by parts twice and using the fact that ψ satisfies (3.1.2) we obtain the following version of Green's identity:

$$\begin{aligned} \int d^4x\, \psi\, (\partial^a\partial_a - m^2)\, Af &= \int_{t \in [t_1, t_2]} d^4x\, \psi(\partial^a\partial_a - m^2)\, Af \\ &= \int_{t_1} d^3x\, [\psi \frac{\partial(Af)}{\partial t} - (Af)\frac{\partial\psi}{\partial t}] \end{aligned} \tag{3.2.26}$$

However, on the hypersurface $t = t_1$, we have $Rf = 0$. Hence, taking account of eq. (3.2.25) and the definition of Ω (see eq. (3.2.3)), we obtain

$$\int d^4x \, \psi f = \Omega(Ef, \psi) \tag{3.2.27}$$

as we desired to show. □

The third property, eq. (3.2.27), of lemma 3.2.1 shows that for each $f \in \mathfrak{T}$, the function $\Omega(Ef,\cdot)$ on $\mathcal{S}$ is equal to what one would obtain by averaging the solution over spacetime, weighted by f. This suggests that in the quantum theory, we should identify the (Schrodinger) operator $\hat{\Omega}(Ef,\cdot)$—defined by eq. (3.2.10)—with the spacetime average of the (Heisenberg) quantum field operator, weighted by f. More precisely, define the operator $\hat{\phi}(f)$ by

$$\hat{\phi}(f) = \hat{\Omega}(Ef,\cdot) = ia(\overline{K(Ef)}) - ia^{\dagger}(K(Ef)) \tag{3.2.28}$$

Then $\hat{\phi}(f)$ has the interpretation of being the spacetime average (weighted by f) of the Heisenberg operator representing the value of the field at point **x** at time t. Indeed, eq. (3.2.28) corresponds to what one would obtain by multiplying the formal expression (3.1.19) by f and integrating over spacetime. Thus, although eq. (3.1.19) is mathematically ill-defined, eq. (3.2.28) provides a mathematically rigorous definition of the Heisenberg field operator $\hat{\phi}$ as an "operator-valued distribution" on spacetime, i.e. as a map from test functions into operators on $\mathfrak{F}_s(\mathcal{H})$.

Note that we may view the collection of all "smeared field operators", defined by eq. (3.2.28), as comprising the fundamental observables of the quantum field theory, since by property (1) of lemma 3.2.1, the specification of these observables for all $f \in \mathfrak{T}$ is equivalent to the specification of the observables $\hat{\Omega}(\psi,\cdot)$ for all $\psi \in \mathcal{S}$. Note also that the field operator $\hat{\phi}$ satisfies the distributional version of the Klein-Gordon equation, eq. (3.1.2), in the sense that for all $g \in \mathfrak{T}$ we have

$$\hat{\phi}\,([\partial^a\partial_a - m^2]g) = \hat{\Omega}(E[\partial^a\partial_a - m^2]g,\cdot) = 0 \tag{3.2.29}$$

where property (2) of lemma 3.2.1 was used. In addition, note that in terms of the "smeared field operators", the fundamental commutation relations are

$$
\begin{aligned}
[\hat{\phi}(f), \hat{\phi}(g)] &= [\hat{\Omega}(Ef, \cdot), \hat{\Omega}(Eg, \cdot)] \\
&= -i\Omega\,(Ef, Eg) \\
&= -iE(f,g) \qquad (3.2.30)
\end{aligned}
$$

where property (3) of lemma 3.2.1 was used and where we have written $E(f,g) \equiv \int fEg\, d^4x$. Equation (3.2.30) frequently is expressed more loosely as

$$[\hat{\phi}(x), \hat{\phi}(x')] = -\,iE(x,x') \qquad (3.2.31)$$

Finally, note that it follows immediately from eq. (3.2.28) that

$$
\begin{aligned}
<0|\hat{\phi}(f)\hat{\phi}(g)|0> &= (KEf, KEg)_{\mathcal{H}} \\
&= \mu(Ef,Eg) - \frac{i}{2}\, E(f,g) \qquad (3.2.32)
\end{aligned}
$$

where eq. (3.2.19) was used in the second line. Thus, the bilinear map $\mu: \mathcal{S} \times \mathcal{S} \to \mathbb{R}$ used to construct the quantum field theory is directly and simply related to the real part of the "two-point function", $<0|\hat{\phi}(x)\hat{\phi}(x')|0>$, of the quantum field in the vacuum state defined by the construction.

3.3 Particle Interpretation

In the previous two sections, we presented the quantum theory of a Klein-Gordon field ϕ in Minkowski spacetime without ever mentioning the word "particle". This contrasts sharply with most presentations of quantum field theory, wherein one attempts at the outset to obtain a quantum theory of a relativistic particle, and ends up constructing a quantum field theory only because the theory of a relativistic particle runs into some mathematical difficulties. In such presentations, a particle interpretation of the states of the quantum field usually is taken for granted, since the Hilbert space, $\mathcal{H}$, makes its first appearance in the attempt to define a Hilbert space for a single relativistic particle, and the Fock space, $\mathcal{F}_s(\mathcal{H})$, initially appears in the attempt to allow multiparticle states. However, irrespective of the manner in which the formulation of quantum field theory is introduced, the interpretation of the states of a quantum field in terms of "particles" requires justification.

As our presentation makes manifest, quantum field theory is the quantum theory of a field, not a theory of "particles". However, when we consider the manner in which a quantum field interacts with other systems to which it is coupled, an interpretation of the states in $\mathcal{F}_s(\mathcal{H})$ in terms of "particles" naturally arises. It is, of course, essential that this be the case if quantum field theory is to describe observed phenomena, since "particle-like" behavior is commonly observed. Indeed, experiments designed to explore quantum field phenomena are usually referred to as "particle physics" experiments.

In this section, we shall illustrate how a particle interpretation of states in the standard quantum field theory in Minkowski spacetime arises by analyzing a simple model coupling of a quantum field to a two-level quantum mechanical system (Unruh and Wald 1984). This analysis provides a good illustration of the nature—and limitations—of the notion of "particles" in quantum field theory in flat spacetime. The limitations of this notion are considerably greater in curved spacetime, where a useful particle interpretation of states does not, in general, exist.

Consider a two level quantum mechanical system which is linearly coupled to a Klein-Gordon quantum field, ϕ, in Minkowski spacetime. More precisely, we consider a combined system whose total Hamiltonian is time-independent and is of the form

$$H = H_\phi + H_q + H_{int} \tag{3.3.1}$$

Here, H_ϕ denotes the Hamiltonian of the free Klein-Gordon field theory constructed in the previous section. The unperturbed two-level quantum system is taken to have energy eigenstates $|\mathfrak{X}_0\rangle$ and $|\mathfrak{X}_1\rangle$, with energies 0 and σ respectively, so its (unperturbed) Hamiltonian is taken to have the form

$$H_q = \sigma A^\dagger A \tag{3.3.2}$$

where the operator A is defined by

$$A|\mathfrak{X}_0\rangle = 0, \quad A|\mathfrak{X}_1\rangle = |\mathfrak{X}_0\rangle \tag{3.3.3}$$

We choose the interaction Hamiltonian, H_{int}, of this model to be

$$H_{int} = \varepsilon(t) \int d^3x \; \hat{\phi}(\mathbf{x}) \, [F(\mathbf{x})A + \text{h.c.}] \tag{3.3.4}$$

where the spatial function $F(\mathbf{x})$ is in $C_0^\infty(\mathbb{R}^3)$, and the coupling constant $\varepsilon(t)$ is smoothly "turned on" and "turned off" at finite times, so that $\varepsilon \in C_0^\infty(\mathbb{R})$.

We wish to calculate—to lowest order in ε—the transitions of the two-level system induced by its coupling to the field, as well as the resulting changes in the state of the field. As in most calculations in time-dependent perturbation theory, it is convenient to work in the "interaction picture". From the definition, (3.3.3), of A, it follows immediately that in the interaction picture, we have

$$A_I(t) = e^{-i\sigma t} A_s \tag{3.3.5}$$

where A_s denotes the (time-independent) Schrodinger picture operator. Hence, in the interaction representation, the interaction Hamiltonian can be expressed as

$$(H_{int})_I = \int d^3x\ [\varepsilon(t)\, e^{-i\sigma t}\, F(\mathbf{x})\, \phi_I(t, \mathbf{x})\, A_s + \text{h.c.}] \tag{3.3.6}$$

We consider the case where initially (i.e., before the interaction is "turned on"), the two-level system is in an arbitrary state $|\mathfrak{X}>$ and the field is in the state

$$|n_\psi> \equiv (0,.., 0, \psi^{a_1} \ldots \psi^{a_n}, 0,\ldots) \tag{3.3.7}$$

for some $\psi \in \mathcal{H}$ with unit norm, where the Fock space index notation used here is explained in Appendix A.3. (In standard terminology, the state (3.3.7) normally would be called an "n-particle state", with all the individual particles in state ψ. However, I shall refrain from using such terminology at this stage, since the goal of the analysis here is precisely to investigate to what extent this terminology is justified.) Thus, we assume that the initial state of the full system is

$$|\Psi_i> = |\mathfrak{X}>|n_\psi> \tag{3.3.8}$$

Then, from the standard formulas of time-dependent perturbation theory, we find that to first order in ε, the final joint state of the two-level and field system is

$$|\Psi_f> = \left[I - i \int_{-\infty}^{\infty} dt\ (H_{int})_I\right] |\Psi_i> \tag{3.3.9}$$

From eq. (3.3.6), we may write

$$\int_{-\infty}^{\infty} dt\ (H_{int})_I = \phi_I(f)A_s + \text{h.c.} \tag{3.3.10}$$

where

$$f(t,\mathbf{x}) = \varepsilon(t)\, e^{-i\sigma t}\, F(\mathbf{x}) \tag{3.3.11}$$

so that f is a (complex) test function, and by $\phi_I(f)$ we mean,

$$\phi_I(f) = \phi_I(\text{Re } f) + i\, \phi_I(\text{Im } f)\ . \tag{3.3.12}$$

Since the interaction representation coincides with the Heisenberg representation for the noninteracting theory, from eq. (3.2.28) we obtain

$$\phi_I(f) = ia(\overline{KE\bar{f}}) - ia^\dagger(KEf) \tag{3.3.13}$$

where we have extended the action of K from $\mathcal{S}$ to $\mathcal{S}^{\mathbb{C}}$ by complex linearity. Now, if the interaction is "turned on" and "turned off" very slowly—i.e., if ε is a very slowly varying function of time compared with the frequency σ—then f will very nearly be purely positive frequency (see eq. (3.3.11)). In that case, Ef also will be very nearly purely positive frequency, and we obtain

$$KEf \approx Ef,\quad KE\bar{f} \approx 0 \tag{3.3.14}$$

Hence, writing

$$\lambda = -KEf \tag{3.3.15}$$

we find that in the case where the interaction is turned on and off sufficiently slowly, we have

$$\phi_I(f) \approx ia^\dagger(\lambda) \tag{3.3.16}$$

(Note that our definition of λ is such that $||\lambda||$ is proportional to ε.) Thus, combining eqs. (3.3.9), (3.3.10), and (3.3.16), we find that the final state is

$$
\begin{aligned}
|\Psi_f\rangle &= (I + a^\dagger(\lambda)A - a(\bar{\lambda})A^\dagger)\,|\Psi_i\rangle \\
&= |n_\psi\rangle|\mathcal{X}\rangle + \sqrt{n+1}\,\|\lambda\|\,(A|\mathcal{X}\rangle)\,|(n+1)'\rangle \\
&\quad - \sqrt{n}\,(\lambda,\psi)\,(A^\dagger|\mathcal{X}\rangle)\,|(n-1)_\psi\rangle \qquad (3.3.17)
\end{aligned}
$$

where

$$|(n+1)'\rangle \equiv (0,\ldots,\, 0, \psi^{(a_1}\ldots\psi^{a_n}\lambda^{a_{n+1})}/\|\lambda\|,\, 0,\ldots) \qquad (3.3.18)$$

Now consider the case where the quantum mechanical system initially is in its ground state, i.e. $|\mathcal{X}\rangle = |\mathcal{X}_0\rangle$, so that $A|\mathcal{X}\rangle = 0$. Equation (3.3.17) shows that as a result of its interaction with the field, the two-level system may make a transition to its excited state. Furthermore, eq. (3.3.17) shows that—to lowest order in perturbation theory—this transition always is accompanied by a transition of the field from state $|n_\psi\rangle$ to state $|(n-1)_\psi\rangle$. The probability that this joint transition occurs is proportional to n.

Similarly, we find that if the quantum mechanical system initially is in its excited state $|\mathcal{X}_1\rangle$, a transition to its ground state can occur. To lowest order, this transition always is accompanied by a transition of the field to the state $|(n+1)'\rangle$, and, in the simple case in which λ is proportional to ψ, the probability of making such a transition is proportional to $(n+1)$.

A "particle language" can be usefully employed to describe the results summarized in the previous two paragraphs. We interpret any Fock space state with a nonzero component only in $\overset{n}{\otimes}_s\mathcal{H}$ as corresponding to the presence of precisely n "particles" (or "quanta") of the field ϕ. In particular, the initial state $|n_\psi\rangle$ we have chosen for the field is interpreted as describing n particles, each in state ψ. The transition $|n_\psi\rangle|\mathcal{X}_0\rangle \to |(n-1)_\psi\rangle|\mathcal{X}_1\rangle$ then can be interpreted as corresponding to the "absorption of a particle". This fits in well with the fact that this transition probability is proportional to n. In a similar manner, the transition $|n_\psi\rangle|\mathcal{X}_1\rangle \to |(n+1)'\rangle|\mathcal{X}_0\rangle$ can be interpreted as describing the emission of a particle in mode $\lambda/\|\lambda\|$.

Thus, the behavior of the quantum field interacting with the quantum mechanical system is modeled quite well by the following simple picture: One views the quantum field theory as actually being a theory of "particles", with the states of the field given the particle interpretation described above. One views the interactions of the field with the quantum mechanical system as enabling the quantum mechanical system to absorb or emit these particles—making appropriate energy conserving transitions of its state when it does so.

This naive particle interpretation of the theory does a remarkably good job of modeling the main features of the interactions of the quantum field with other systems. Note, however, that even in the context of the above model, there are some features which are not well described by the naive particle picture. In particular, as already noted, the probability of making a downward transition is proportional to $(n+1)$. The fact that this probability is nonvanishing even when $n = 0$ is normally described in the particle interpretation by saying that the quantum mechanical system has the ability to "spontaneously" emit a particle. To explain the enhanced probability of making a downward transition when particles are initially present, one invents some new "intuition" and language; the term "stimulated emission" is used to describe this particular effect.

Probably the most misleading aspect of the particle viewpoint is the promotion of a naive view of the vacuum state as an entirely inert entity. The statement that "no particles are present" strongly suggests that the quantum field exerts no influence on other systems when it is in the vacuum state. The fact that it does is then further hidden by the use of terminology like the above-mentioned "spontaneous" emission. In fact, as our calculation explicitly demonstrates, it is the interaction of the quantum mechanical system with the quantum field occurring when the field initially is in its vacuum state which is entirely responsible for the phenomenon of "spontaneous" emission by the quantum mechanical system. The misleading picture of the vacuum state promoted by the particle interpretation is probably the greatest barrier which must be overcome in order to understand the Unruh effect (see chapter 5).

Nevertheless, the particle interpretation is quite useful for describing many phenomena in quantum field theory in flat spacetime; and I shall freely make use of this language below. In particular, henceforth I shall refer to the Hilbert space $\mathcal{H}$ occurring in the construction of the quantum field theory as the "one-particle Hilbert space". However, it always should be borne in mind that the notion of "particles"—while quite useful in certain contexts—plays no fundamental role in the formulation of quantum field theory. As we shall see in section 4.3 below, in a curved stationary spacetime, a similar particle interpretation of quantum field theory also can be given. In a curved spacetime which is asymptotically stationary in the past or future, a natural particle interpretation of states can be given in these asymptotic regions. However, in a general, curved spacetime, it does not appear that any natural notion of particles exists. Nevertheless, as we shall see explicitly in the next chapter, the lack of a useful particle interpretation of quantum field theory in a general curved spacetime does not prevent one from giving a completely satisfactory formulation of the theory.

4 Quantum Fields in Curved Spacetime

In the previous chapter, we formulated quantum field theory in Minkowski spacetime in a way that eliminated any reference to a "plane wave basis". Indeed, the construction given in section 3.2 above relied upon only the structure naturally present in the classical theory, together with the choice of $\mathcal{H}$ as the space of "positive frequency solutions". We also described how this construction could be generalized to allow other suitable choices of $\mathcal{H}$. Consequently, we now are in a position to formulate quantum field theory in curved spacetime by a straightforward adaptation of this prescription. However, in order to do so, it is necessary that the causal behavior of the curved spacetime be sufficiently well behaved that the space of solutions to the classical field equations have the same basic structure as in Minkowski spacetime. As we shall see in section 4.1, the condition of global hyperbolicity ensures that this is the case. We shall construct quantum field theory in curved, globally hyperbolic spacetimes in section 4.2. The only significant difference from the Minkowski spacetime case is that, in general, we now do not have a preferred choice of $\mathcal{H}$, so we must consider (on an essentially equal footing) the entire class of constructions. One important exception to this statement occurs for the special case of stationary spacetimes, where—as we shall see in section 4.3—a natural prescription exists for defining $\mathcal{H}$ as the space of "positive frequency solutions". Section 4.4 analyzes the unitary equivalence of the different constructions and—as a very important byproduct—obtains an explicit construction of the S-matrix describing particle creation and scattering in situations where asymptotic notions of particles exist. The fact that different constructions can yield unitarily inequivalent quantum field theories is then faced directly in section 4.5. It is shown there how the algebraic approach allows one to give a fully satisfactory formulation of quantum field theory in curved spacetime which encompasses all of the different, unitarily inequivalent constructions. However, the theory is still a "minimal" one in that only the primary field observables, $\hat{\Omega}(\psi,\cdot)$, are directly represented. An additional observable of great importance for determining the "back-reaction" effects of the quantum field on the curved spacetime is the stress-energy tensor of the field. In section 4.6, we show how the expected stress-energy of states may be determined (up to some "local curvature" ambiguity), and we discuss the back-reaction problem. The Hadamard condition on states is introduced in section 4.6 as a necessary condition for a state to have a finite expected stress-energy, and it also is noted that the imposition of the Hadamard condition greatly restricts the number of unitarily inequivalent Hilbert space constructions of the theory. The chapter concludes with a brief discussion in section 4.7 of the formulation of quantum field theory in curved spacetime for other linear fields.

4.1 Curved Spacetimes; Global Hyperbolicity

In the framework of general relativity, spacetime structure is described by a four-dimensional manifold, M, on which there is present a Lorentz metric, g_{ab}. In addition, in general relativity g_{ab} is related to the matter distribution by Einstein's equation. However, except for issues related to "back-reaction" (to be discussed in section 4.6), the dynamics of g_{ab} will not concern us here. Thus, in the following, we shall view the spacetime structure (M, g_{ab}) as being given, and focus upon the formulation of the quantum theory of a Klein-Gordon scalar field, ϕ, in this fixed, curved background. Unless otherwise stated, we shall assume throughout these notes that g_{ab} is smooth.

The classical action of a Klein-Gordon field in curved spacetime is

$$S = -\frac{1}{2}\int (\nabla_a\phi\nabla^a\phi + m^2\phi^2)\sqrt{-g}\, d^4x \qquad (4.1.1)$$

and the curved spacetime version of the Klein-Gordon equation thus is

$$\nabla^a\nabla_a\phi - m^2\phi = 0 \qquad (4.1.2)$$

where ∇_a is the derivative operator compatible with g_{ab}. Other natural generalizations of eq. (3.1.2) to curved spacetime involving the addition of a term $\xi R\phi$ to eq. (4.1.2) are possible (where ξ is a constant and R denotes the scalar curvature). All of our discussion below would apply with equal validity to such modified versions of eq.(4.1.2).

In an arbitrary curved spacetime, the classical existence and uniqueness properties of solutions to eq. (4.1.2) can be very different from that of Minkowski spacetime. Two examples will suffice to illustrate this point: (i) Let spacetime be a flat 4-torus, with spatial periodicity L and time periodicity T, with T^2/L^2 irrational. Then in this spacetime, eq. (4.1.2) with m = 0 admits only the solution ϕ = constant. (ii) Consider any spacetime with a "timelike singularity", such as Minkowski spacetime with a timelike line removed, or the Schwarzschild solution with negative mass. Since eq. (4.1.2) does not restrict what can emerge from a singularity, there is no possibility that uniqueness can hold for solutions to eq. (4.1.2) with given initial conditions on a spacelike hypersurface.

We wish to restrict attention here to spacetimes where the classical dynamics governed by eq. (4.1.2) has a well posed initial value formulation analogous to that of Minkowski spacetime. In such cases, we shall be able to formulate quantum field theory constructions in a manner completely analogous to our construction of the previous chapter for a general choice of $\mathcal{H}$. Generalizations of quantum field theory to spacetimes which do not admit a well posed classical initial value formulation may then be considered (see, in particular, Kay (1992)), although we shall not do so in these notes.

Fortunately, there is a simple condition on a spacetime (M, g_{ab}) which guarantees that equations of the general form (4.1.2) have a well posed initial value formulation. First, we restrict attention to spacetimes which are time orientable, i.e., such that a continuous choice can be made throughout the spacetime of which half of each light cone constitutes the "future" direction and which half constitutes the "past." Let $\Sigma \subset M$ be any closed set which is achronal, i.e., no pair of points $p,q \in \Sigma$ can be joined by a timelike curve. We define the *domain of dependence* of Σ by

$$D(\Sigma) = \{ p \in M \mid \text{every (past and future) inextendible causal curve through } p \text{ intersects } \Sigma \} \qquad (4.1.3)$$

(See, e.g., Wald (1984a) for the definition of extendibility of causal curves and further discussion of issues relating to the causal structure of spacetimes.) If $D(\Sigma) = M$, then Σ is said to be a *Cauchy surface* for the spacetime (M, g_{ab}). It then follows that a Cauchy surface is automatically a 3-dimensional, C^0, hypersurface. If a spacetime admits a Cauchy surface, then the spacetime is said to be *globally hyperbolic.*

A key theorem regarding the structure of globally hyperbolic spacetimes is the following:

Theorem 4.1.1 (Geroch 1970; Dieckmann 1988) : If (M, g_{ab}) is globally hyperbolic with Cauchy surface Σ, then M has topology $\mathbb{R} \times \Sigma$. Furthermore, M can be foliated by a one-parameter family of smooth Cauchy surfaces Σ_t, i.e., a smooth "time coordinate" t can be chosen on M such that each surface of constant t is a Cauchy surface.

Since every causal curve "registers" on a Cauchy surface Σ, one might expect to have a well defined, deterministic classical evolution from initial conditions given on Σ. That this is indeed the case for eq. (4.1.2) is stated in the following theorem (see, e.g., Hawking and Ellis (1973)):

Theorem 4.1.2: Let (M, g_{ab}) be a globally hyperbolic spacetime with smooth, spacelike Cauchy surface Σ. Then the Klein-Gordon equation (4.1.2) has a well posed initial value formulation in the following sense: Given any pair of smooth (C^∞) functions (ϕ_0, $\dot{\phi}_0$), on Σ, there exists a unique solution, ϕ, to (4.1.2), defined on all of M, such that on Σ we have $\phi = \phi_0$ and $n^a \nabla_a \phi = \dot{\phi}_0$, where n^a denotes the unit (future-directed) normal to Σ. Furthermore, for any closed subset $S \subset \Sigma$, the solution, ϕ, restricted to $D(S)$ depends only upon the initial data on S. In addition, ϕ is smooth and varies continuously with the initial data (with a suitable, Sobolev space topology defined on the initial data).

Similar results hold for a much more general class of linear wave equations and systems of linear wave equations.

The above theorem continues to hold if a (fixed, smooth) "source term", f, is inserted on the right side of eq. (4.1.2). It then follows directly from the domain of dependence property of this theorem that there exist unique advanced and retarded solutions to the Klein-Gordon equation with source, i.e., in a globally hyperbolic spacetime there exist unique advanced and retarded Green's functions for the Klein-Gordon equation.

In the following, we shall restrict consideration to globally hyperbolic spacetimes.

4.2 Construction of Quantum Field Theory in a Curved Spacetime

Let (M, g_{ab}) be a globally hyperbolic spacetime. To obtain a phase space formulation of classical dynamical evolution for the Klein-Gordon field, we first introduce a "slicing" of M by spacelike Cauchy surfaces Σ_t, labeled by the parameter t. We then introduce a "time evolution" vector field, t^a, on M, satisfying $t^a \nabla_a t = 1$. We decompose t^a as

$$t^a = Nn^a + N^a \tag{4.2.1}$$

where n^a is the unit normal to Σ_t and N^a is tangential to Σ_t. (We refer to N as the *lapse function* and N^a as the *shift vector.*) We introduce local coordinates t, x^1, x^2, x^3 with $t^a \nabla_a x^\alpha = 0$ for $\alpha = 1,2,3$, so that $t^a = (\partial/\partial t)^a$. The Klein-Gordon action (4.1.1) then takes the form

$$S = \int \mathcal{L}\, dt \tag{4.2.2}$$

with

$$\mathcal{L} = \frac{1}{2}\int_{\Sigma_t}[(n^a \nabla_a \phi)^2 - h^{ab}\nabla_a \phi \nabla_b \phi - m^2 \phi^2]\, N \sqrt{h}\, d^3x \tag{4.2.3}$$

where h_{ab} is the induced (Riemannian) metric on Σ_t.

Using the relation

$$n^a \nabla_a \phi = \frac{1}{N}(t^a - N^a)\nabla_a \phi = \frac{1}{N}\dot{\phi} - \frac{1}{N} N^a \nabla_a \phi \tag{4.2.4}$$

we find that the momentum density, π, canonically conjugate to the configuration variable ϕ on Σ_t is given by

$$\pi = \frac{\delta S}{\delta \dot{\phi}} = (n^a \nabla_a \phi)\sqrt{h} \tag{4.2.5}$$

Thus, a point in the classical phase space, $\mathcal{M}$, of Klein-Gordon theory in curved spacetime consists of the specification of functions $\phi(x)$ and $\pi(x)$ on a Cauchy surface Σ_0. In order that all structures be mathematically well defined, we specify $\mathcal{M}$ precisely by again requiring that ϕ and π be smooth and of compact support. Again, we define $\mathcal{S}$ to be the space of solutions to eq. (4.1.2) which arise from initial data in $\mathcal{M}$. By theorem 4.1.2 above, each $[\phi,\pi]\in\mathcal{M}$, gives rise to a unique element of $\mathcal{S}$, so we may again identify $\mathcal{M}$ and $\mathcal{S}$. In addition, theorem 4.1.2 implies that $\mathcal{S}$ is independent of the choice of Σ_0, since $[\phi,\pi]$ will be smooth and of compact support on the Cauchy surface Σ_0 if and only if it is smooth and of compact support on all Cauchy surfaces. Note that dynamical evolution in $\mathcal{M}$ by "time" t corresponds to evaluation of the associated element of $\mathcal{S}$ (and the canonically conjugate momentum π of this solution) on the Cauchy surface Σ_t.

The symplectic structure, Ω, on $\mathcal{M}$ is given by

$$\begin{aligned}\Omega([\phi_1,\pi_1],[\phi_2,\pi_2]) &= \int_{\Sigma_0}(\pi_1\phi_2-\pi_2\phi_1)\, d^3x \\ &= \int_{\Sigma_0}[\phi_2\, n^a\nabla_a\phi_1-\phi_1\, n^a\nabla_a\phi_2]\sqrt{h}\, d^3x \qquad (4.2.6)\end{aligned}$$

Again, Ω is conserved for solutions, so we may view Ω as a bilinear map on $\mathcal{S}$, i.e., $\Omega:\mathcal{S}\times\mathcal{S}\to\mathbb{R}$.

As in Minkowski spacetime, we define the space, $\mathcal{T}$, of "test functions" on a globally hyperbolic curved spacetime, (M, g_{ab}), to be $\mathcal{T} = C_0^\infty(M)$. The existence of unique advanced and retarded Green's functions again allows us to define the map $E:\mathcal{T}\to\mathcal{S}$ by eq. (3.2.22). The proofs of the three properties of this map expressed in lemma 3.2.1 then generalize straightforwardly to curved spacetime. Thus, all of the mathematical structure used to formulate the quantum theory of a Klein-Gordon field in Minkowski spacetime carries over to globally hyperbolic curved spacetimes with one major exception: In a general, curved spacetime, there is no unique, natural analog of the space of "positive frequency solutions" used to define $\mathcal{H}$.

Nevertheless, there is no difficulty in constructing a quantum field theory for the Klein-Gordon field as follows: Choose any bilinear map $\mu:\mathcal{S}\times\mathcal{S}\to\mathbb{R}$ such that for all $\psi_1\in\mathcal{S}$,

$$\mu(\psi_1,\psi_1) = \frac{1}{4}\ \underset{\psi_2\neq 0}{\text{l.u.b.}}\ \frac{[\Omega(\psi_1,\psi_2)]^2}{\mu(\psi_2,\psi_2)} \qquad (4.2.7)$$

Construct the Hilbert space $\mathcal{H}$ and the projection map $K:\mathcal{S}\to\mathcal{H}$ by the algorithm given in section 3.2. Define the quantum field theory by taking the Hilbert space to be $\mathcal{F}_s(\mathcal{H})$ and representing the classical observables $\Omega(\psi,\cdot)$ for each $\psi\in\mathcal{S}$ by the operator

$$\hat{\Omega}(\psi,\cdot) = ia(\overline{K\psi}) - ia^{\dagger}(K\psi) \tag{4.2.8}$$

(see eq. (3.2.10)). Equivalently, for each test function $f \in \mathcal{T}$, define the "smeared Heisenberg field operator" $\hat{\phi}(f) : \mathcal{F}_s(\mathcal{H}) \rightarrow \mathcal{F}_s(\mathcal{H})$ by

$$\hat{\phi}(f) = ia(\overline{K(Ef)}) - ia^{\dagger}(K(Ef)) \tag{4.2.9}$$

The only unsatisfactory aspect of this construction is that it involves an arbitrary (i.e., unspecified) choice of the bilinear map μ. Although, as discussed at the end of the next section, there never is any difficulty with the existence of μ's which satisfy eq. (4.2.7), in a general, curved spacetime there does not appear to be any "preferred" choice of μ. There are two important ramifications of this fact.

First, since the choice of μ is in one-to-one correspondence with the construction of the Hilbert space of states, $\mathcal{F}_s(\mathcal{H})$, and since the Fock space structure of $\mathcal{F}_s(\mathcal{H})$, in turn, allows one to define a notion of "particles", this means that there is no natural definition of "particles" in a general, curved spacetime. Indeed, although a considerable amount of effort has been expended by researchers to obtain a preferred definition of "particles", these efforts have not been successful, except in some restricted classes of spacetimes (most notably, the class of stationary spacetimes discussed in the next section) where symmetries or other structure can be used to naturally select a μ. To put this statement in perspective, we should point out that there is no difficulty defining an approximate notion of "particles" when the spacetime is nearly flat (or, more generally, nearly stationary); more precisely, the notion of "particles" becomes highly ambiguous only for modes whose frequency is smaller than typical inverse timescales for the change of the metric. Thus, in our universe, serious difficulties would arise if we attempt to define a meaningful notion of "particles" whose wavelength is larger than the Hubble radius, but, as a practical matter, there is little difficulty employing a particle concept in most other circumstances. Nevertheless, as a matter of principle, it should be stressed that, in general, the notion of "particles" in curved spacetime is, at best, only an approximate one. While some readers familiar with standard presentations of quantum field theory in flat spacetime might be disturbed by the lack of a notion of "particles" in curved spacetime, we have taken great care to emphasize here that this should not be a cause of alarm, since the notion of "particles" plays no essential role in the formulation of quantum field theory. Indeed, I view the lack of an algorithm for defining a preferred notion of "particles" in quantum field theory in curved spacetime to be closely analogous to the lack of an algorithm for defining a preferred system of coordinates in classical general relativity. (Readers familiar only with presentations of special relativity based upon the use of global inertial coordinates might well find this fact to be alarming.) In both cases, the lack of an algorithm does not, by itself, pose any difficulty for the formulation of the theory.

However, a second, related ramification of the lack of a preferred choice of μ does pose a potentially serious difficulty with the formulation of quantum field theory in curved spacetime. If the Stone-von Neuman theorem (see theorem 2.2.1 above) were valid for a quantum theory corresponding to an infinite dimensional phase space, $\mathcal{M}$, then the choice of μ would be of no consequence, since different choices of μ would lead to unitarily equivalent theories. In that case, the choice of μ would be exactly analogous to the choice of a coordinate system in classical general relativity in that, in both cases, the choice might affect the superficial appearance of some formulas but would not affect any physical predictions. However, in section 4.4 below, we shall see explicitly that when $\mathcal{M}$ is infinite dimensional, different choices of μ can yield unitarily inequivalent theories. Thus, to define quantum field theory in a general curved spacetime, it might appear to be essential that we prescribe a unique unitary equivalence class of μ's. In fact, we shall see in section 4.6 that the imposition of the Hadamard condition on states does define a preferred unitary equivalence class of μ's in the case of a "closed universe" (i.e., a spacetime with compact Cauchy surface). However, it does not appear that a uniquely preferred unitary equivalence class of μ's exists for general spacetimes with a non-compact Cauchy surface. Fortunately, we shall see in section 4.5 below that this fact also does not pose a serious problem: Quantum field theory in curved spacetime can be formulated via the algebraic approach in a manner which does not require the specification of even a preferred unitary equivalence class of μ's.

Despite the lack of a completely general algorithm to define μ, in stationary spacetimes (i.e., spacetimes possessing a time translation symmetry) there does exist a natural, unique choice of μ.

Furthermore, the notion of "particles" associated with this μ has a direct physical interpretation. We now shall describe this quantum field theory construction for stationary spacetimes.

4.3 Quantum Field Theory in Stationary Spacetimes

Let (M, g_{ab}) be a globally hyperbolic spacetime which is stationary, i.e., (M, g_{ab}) admits a one-parameter group of isometries, α_t: $M \to M$, whose orbits are timelike. Let ξ^a denote the Killing vector field which generates these isometries. The basic idea of the quantum field construction in stationary spacetimes is to choose $\mathcal{H}$ to be the subspace of (complex) solutions which are "purely positive frequency" with respect to "Killing time" t (satisfying $\xi^a \nabla_a t = 1$) when Fourier analyzed along the orbits of ξ^a. However, if one attempts to straightforwardly implement this construction in a naive manner, one encounters the difficulties that, *a priori,* it is not clear that solutions decay suitably rapidly for their Fourier transform along individual Killing orbits to exist, or that there are "sufficiently many" solutions that are purely positive frequency along all Killing orbits. Nevertheless, a mathematically precise implementation of this idea has been given by Ashtekar and Magnon (1975) and Kay (1978), and we now describe this construction.

First, in order to avoid possible "infra-red divergence" difficulties in the construction (see below), in general we must make the additional restrictions that the mass, m, of the Klein-Gordon field be strictly positive:

$$m > 0 \tag{4.3.1}$$

and that there exists a Cauchy surface, Σ, such that on Σ we have, for some $\varepsilon > 0$,

$$-\xi^a \xi_a \geqq -\varepsilon\, \xi^a n_a > \varepsilon^2 \tag{4.3.2}$$

where n^a denotes the unit normal to Σ. Without the restrictions (4.3.1) and (4.3.2), the quantum field construction given below may still work, but it is not guaranteed to do so.

We complexify $\mathcal{S}$ and define an "energy inner product" on $\mathcal{S}^{\mathbb{C}}$ by

$$<\psi_1, \psi_2> = \int_\Sigma T_{ab}\, \xi^a n^b \sqrt{h}\, d^3x \tag{4.3.3}$$

where the classical stress-energy tensor, T_{ab}, on $\mathcal{S}^{\mathbb{C}}$ is given by

$$T_{ab}(\psi_1, \psi_2) = \nabla_{(a}\overline{\psi}_1 \nabla_{b)}\psi_2 - \frac{1}{2} g_{ab}[\nabla^c\overline{\psi}_1 \nabla_c\psi_2 + m^2\overline{\psi}_1\psi_2] \qquad (4.3.4)$$

(Note that the assumption that m is positive is used here to ensure that <, > is positive definite even if Σ is compact.) Since ψ_1 and ψ_2 are solutions of the Klein-Gordon equation, we have $\nabla^a T_{ab} = 0$. Since $T_{ab} = T_{ba}$ and $\nabla_{(a}\xi_{b)} = 0$ by Killing's equation, it follows that $\nabla^a(T_{ab}\xi^b) = 0$. Hence, using Gauss' law, we see that <, > is independent of the choice of Cauchy surface Σ. In particular, it follows that <, > is invariant under the time translation map τ_t: $\mathcal{S}^{\mathbb{C}} \to \mathcal{S}^{\mathbb{C}}$ defined by $\tau_t(\psi) = \psi \circ \alpha_{-t}$, since applying τ_t to the solutions is equivalent to applying α_{-t} to Σ.

Next, we complete $\mathcal{S}^{\mathbb{C}}$ in the norm defined by <, > to get a complex Hilbert space $\tilde{\mathcal{H}}$. (It should be emphasized that $\tilde{\mathcal{H}}$ is *not* the Hilbert space we seek; it is used only as an intermediate tool in the construction.) The time translation map τ_t extends to $\tilde{\mathcal{H}}$ to define a strongly continuous, one-parameter, unitary group, V_t. By Stone's theorem, V_t is of the form $V_t = \exp(-i\tilde{h}t)$, where $\tilde{h}$: $\tilde{\mathcal{H}} \to \tilde{\mathcal{H}}$ is self-adjoint. Note that from the definition of τ_t, we have for all $\psi \in \mathcal{S}^{\mathbb{C}}$,

$$\tilde{h}\,\psi = i\mathcal{L}_\xi\psi \qquad (4.3.5)$$

where $\mathcal{L}$ denotes the Lie derivative.

Now, define B: $\mathcal{S}^{\mathbb{C}} \times \mathcal{S}^{\mathbb{C}} \to \mathbb{C}$ by

$$B(\psi_1, \psi_2) = \Omega(\overline{\psi}_1, \psi_2) \qquad (4.3.6)$$

Using the definition of Ω, eq.(4.2.6), together with eqs. (4.3.1), (4.3.2), and the Schwarz inequality, we obtain

$$|B(\psi_1, \psi_2)| \leq C\,\|\psi_1\|\,\|\psi_2\| \qquad (4.3.7)$$

from which it follows that B extends continuously to a quadratic form on $\tilde{\mathcal{H}}$. Furthermore, by a direct calculation using eq. (4.3.5) and the definitions of Ω and <, >, it may be verified that for all $\psi_1, \psi_2 \in \mathcal{S}^{\mathbb{C}}$, we have

$$B(\psi_1, \tilde{h}\psi_2) = 2\,i < \psi_1, \psi_2 > \qquad (4.3.8)$$

From eqs. (4.3.7) and (4.3.8) and the fact that $\mathcal{S}^{\mathbb{C}}$ is dense in $\tilde{\mathcal{H}}$, it follows that the spectrum of $\tilde{h}$ is bounded away from zero, so $\tilde{h}^{-1}$ exists as a bounded operator on $\tilde{\mathcal{H}}$.

Now, let $\tilde{\mathcal{H}}^+$ be the positive spectral subspace of $\tilde{h}$ in $\tilde{\mathcal{H}}$ (defined via the spectral theorem), so that—in view of eq. (4.3.5)—$\tilde{\mathcal{H}}^+$ may be interpreted as the subspace of "positive frequency solutions" in $\tilde{\mathcal{H}}$. Let $K:\tilde{\mathcal{H}}\rightarrow\tilde{\mathcal{H}}^+$ denote the projection map onto $\tilde{\mathcal{H}}^+$. For all $\psi_1,\psi_2\in\mathcal{S}$, define $\mu:\mathcal{S}\times\mathcal{S}\rightarrow\mathbb{R}$ by

$$\mu(\psi_1,\psi_2) = \text{Im}\, B(K\psi_1,K\psi_2) = 2\,\text{Re} < K\psi_1,\ \tilde{h}^{-1}K\psi_2 > \qquad (4.3.9)$$

Then μ satisfies eq. (4.2.7) with the least upper bound for ψ_2 in that equation attained by a sequence in $\mathcal{S}$ approaching the vector $i(K\psi_1 - \overline{K\psi_1})$ in $\tilde{\mathcal{H}}$. This defines the desired quantum field theory which naturally arises from the stationary symmetry. The "one-particle-Hilbert space", $\mathcal{H}$, defined by the construction is just the completion of the space, $\tilde{\mathcal{H}}^+$, of "positive frequency solutions" in the Klein-Gordon inner product (see eq. (3.2.19)). Note, however, that the construction avoided any direct attempt to take the "time Fourier transform" of solutions in $\mathcal{S}$ along orbits of ξ^a. As already indicated above, significant technical difficulties would have arisen had we attempted to do so, since the time decay properties of solutions in general spacetimes—in particular, in spacetimes with a compact Cauchy surface—are such that this time Fourier transform may exist only in a distributional sense.

It is important to note that the discussion of the "particle interpretation" of states for quantum field theory in Minkowski spacetime given in section 3.3 carries over without any essential change to globally hyperbolic, stationary, curved spacetimes. Namely, a nearly "time-independent" (with respect to ξ^a) coupling of the quantum field ϕ to a time-independent quantum mechanical system will induce transitions of that system in exactly the same manner as discussed in section 3.3. Consequently, the states in the Hilbert space $\mathcal{F}_s(\mathcal{H})$ constructed above have the same type of particle interpretation as in Minkowski spacetime.

Now, consider a globally hyperbolic spacetime (M, g_{ab}) which is "stationary in the past", i.e., suppose that for some Cauchy surface Σ, the past of Σ is isometric to the past of a Cauchy surface Σ' in some globally hyperbolic stationary spacetime (M', g'_{ab}). Then the solution space, $\mathcal{S}$, for (M, g_{ab}) can be identified with the solution space, $\mathcal{S}'$, for (M', g'_{ab}) by identifying initial data for the solutions on corresponding Cauchy surfaces in the isometric regions. Hence, the μ' defined on $\mathcal{S}'$ by eq. (4.3.9) gives rise to a quantum field construction on (M, g_{ab}). The one-particle Hilbert space for the quantum field on (M, g_{ab}) which is obtained from μ' is denoted as $\mathcal{H}_{in}$. States in the Fock space $\mathcal{F}_s(\mathcal{H}_{in})$ obtained from this construction will have a natural particle interpretation in the past, but not necessarily anywhere else in the spacetime. Similar considerations apply if (M, g_{ab}) is merely "asymptotically stationary" in the past, i.e., if, under an appropriate identification of the given spacetime with a stationary spacetime (M', g'_{ab}), solutions in $\mathcal{S}$ suitably approach solutions in $\mathcal{S}'$ in the asymptotic past.

If (M, g_{ab}) also is stationary (or asymptotically stationary) in the future, a similar quantum field construction yields a Fock space $\mathcal{F}_s(\mathcal{H}_{out})$, whose states have a natural particle interpretation in the future. If this construction is unitarily equivalent to the one arising from $\mathcal{H}_{in}$, the unitary map $U:\mathcal{F}_s(\mathcal{H}_{in})\rightarrow\mathcal{F}_s(\mathcal{H}_{out})$ which implements this equivalence is known as the *S-matrix*. Thus, the S-matrix relates the particle description of states in the past with that in the future, and it thereby directly contains all information regarding "spontaneous particle creation" and scattering. We shall compute the S-matrix explicitly (in terms of operators describing classical scattering) in the next section.

Finally, we remark that the considerations of this section allow us to prove that for the Klein-Gordon field on an arbitrary globally hyperbolic spacetime, there always exists a bilinear map $\mu:\mathcal{S}\times\mathcal{S}\rightarrow\mathbb{R}$ satisfying eq. (4.2.7). Namely, given any globally hyperbolic spacetime (M, g_{ab}) with Cauchy surface Σ we can construct a new globally hyperbolic "interpolating spacetime" $(\tilde{M}, \tilde{g}_{ab})$ with Cauchy surfaces $\tilde{\Sigma}_1, \tilde{\Sigma}_2$ such that the future of $\tilde{\Sigma}_1$ in $\tilde{M}$ is isometric to the future of Σ in M, and the past of $\tilde{\Sigma}_2$ is isometric to the past of a Cauchy surface in a stationary, globally hyperbolic spacetime, satisfying eq. (4.3.2) (see Appendix C of Fulling et al (1981)). In addition, in the case of a massless field on (M, g_{ab}) we may also "deform" m on $(\tilde{M}, \tilde{g}_{ab})$ to the past of $\tilde{\Sigma}_1$ (by treating it as an external potential) so that m is nonzero and constant in the past of $\tilde{\Sigma}_2$, in which case eq. (4.3.1) also is satisfied there. By identifying the solution space, $\mathcal{S}$, for (M, g_{ab}) with the solution space $\tilde{\mathcal{S}}$ for $(\tilde{M}, \tilde{g}_{ab})$ and, in turn, identifying $\tilde{\mathcal{S}}$ with the solution space for the stationary spacetime, we thereby obtain the desired μ.

Indeed, a much more direct proof of the existence of a bilinear map $\mu:\mathcal{S}\times\mathcal{S}\to\mathbb{R}$ satisfying eq. (4.2.7) can be given as follows (Chmielowski 1994). Start with any inner product, <, >, on $\mathcal{S}^{\mathbb{C}}$ whose associated norm satisfies the inequality

$$\|\psi_1\|\,\|\psi_2\| \geq \frac{1}{2}\,|\Omega(\psi_1,\psi_2)| \tag{4.3.10}$$

(An example of such an inner product is the sum of the L^2-norms of ψ and its normal derivative on an arbitrary Cauchy surface Σ.) Complete $\mathcal{S}^{\mathbb{C}}$ in this inner product to get a complex Hilbert space $\mathcal{H}'$. Since the quadratic form $i\Omega(\psi_1,\psi_2)$ is bounded on $\mathcal{H}'$, it defines a bounded operator $A:\mathcal{H}'\to\mathcal{H}'$ such that $<\psi_1, A\psi_2> = i\Omega(\overline{\psi}_1,\psi_2)$. Furthermore, the antisymmetry property of Ω implies that A is self-adjoint. We define $\mu:\mathcal{S}\times\mathcal{S}\to\mathbb{R}$ by

$$\mu(\psi_1,\psi_2) = \frac{1}{2} <\psi_1, |A|\psi_2> \tag{4.3.11}$$

for all $\psi_1,\psi_2\in\mathcal{S}$, where $|A|:\mathcal{H}'\to\mathcal{H}'$ denotes the absolute value of A, i.e., the positive square root of $A^\dagger A$. Then, it is not difficult to verify that μ satisfies eq. (4.2.7) (see Chmielowski 1994). Indeed, the above stationary vacuum state construction is equivalent to defining μ by this algorithm, starting with the inner product (4.3.3). When $m > 0$, the μ obtained by this algorithm using the inner product (4.3.3) with ξ^a replaced by n^a defines a *Hamiltonian diagonalization* prescription for defining a vacuum state with respect to any Cauchy surface Σ in any globally hyperbolic spacetime. However, unlike the stationary vacuum state construction, the vacuum state obtained by Hamiltonian diagonalization will, in general, depend upon the choice of Cauchy surface Σ.

The above argument establishes quantum field theory constructions of the type discussed in the previous section always exist in any globally hyperbolic spacetime. However, it is clear that the choice of μ is highly nonunique. We turn now to an analysis of the necessary and sufficient conditions for two different choices of μ to yield unitarily equivalent quantum field theories.

4.4 Unitary Equivalence; the S-matrix

Let $(\mathcal{S}, \Omega)$ be an arbitrary real symplectic vector space, i.e., $\mathcal{S}$ is a vector space over $\mathbb{R}$ and the bilinear map $\Omega: \mathcal{S} \times \mathcal{S} \to \mathbb{R}$ is antisymmetric and nondegenerate. The case of interest for us, of course, is where $\mathcal{S}$ is the space of solutions to the Klein-Gordon equation in a globally hyperbolic spacetime having initial data in C_0^∞, with Ω given by eq. (4.2.6); however, it is worth emphasizing that all the considerations of this section and the next require only the symplectic vector space structure of $(\mathcal{S}, \Omega)$. Let $\mu_1: \mathcal{S} \times \mathcal{S} \to \mathbb{R}$ and $\mu_2: \mathcal{S} \times \mathcal{S} \to \mathbb{R}$ be two bilinear maps satisfying eq. (4.2.7). In the manner described previously, we can use these maps to construct, respectively, the Fock spaces $\mathcal{F}_1 = \mathcal{F}_s(\mathcal{H}_1)$ and $\mathcal{F}_2 = \mathcal{F}_s(\mathcal{H}_2)$ with operators $\hat{\Omega}_1(\psi, \cdot): \mathcal{F}_1 \to \mathcal{F}_1$ and $\hat{\Omega}_2(\psi, \cdot): \mathcal{F}_2 \to \mathcal{F}_2$ defined by eq. (4.2.8). We wish to know under what circumstances there will exist a unitary map $U: \mathcal{F}_1 \to \mathcal{F}_2$ such that for all $\psi \in \mathcal{S}$ we have

$$U \hat{\Omega}_1(\psi, \cdot) U^{-1} = \hat{\Omega}_2(\psi, \cdot) \tag{4.4.1}$$

Furthermore, if such a U does exist, we wish to determine it explicitly.

There is a twofold purpose to our investigation of this question. First, it is important with regard to the formulation of quantum field theory in curved spacetime to understand the necessary and sufficient conditions for two different choices of μ to yield unitarily equivalent theories. In particular, if all choices of μ led to unitarily equivalent theories, the choice of μ would be irrelevant and the discussion of section 4.2 (together with the existence proof for μ given at the end of section 4.3) would completely specify the construction of quantum field theory in curved spacetime. However, as we shall see shortly, this is not the case. Secondly, as discussed near the end of the section 4.3, for a spacetime which is asymptotically stationary in both the past and future, we have two natural choices of μ, namely $\mu_1 = \mu_{in}$ and $\mu_2 = \mu_{out}$. The S-matrix $U: \mathcal{F}_1 \to \mathcal{F}_2$ implementing the unitary equivalence of these constructions—if they are unitarily equivalent—yields all information regarding particle creation and scattering. Thus, it is of direct physical relevance to compute U. The basic analysis of the issue of unitary equivalence was given in Shale (1962). Our treatment below will follow closely that of Wald (1979a).

The analysis of unitary equivalence naturally divides into two cases:

<u>case (i)</u>: There exist C, C' > O such that for all $\psi \in \mathcal{S}$ we have

$$C\,\mu_1(\psi,\psi) \leqq \mu_2(\psi,\psi) \leqq C'\mu_1(\psi,\psi) \qquad (4.4.2)$$

<u>case (ii)</u>: No such C, C'> O exist.

We claim, first, that unitary equivalence cannot hold in case (ii). Namely, if, say, there does not exist a C > O such that $\mu_1(\psi,\psi) \leqq \mu_2(\psi,\psi)/C$ for all $\psi \in \mathcal{S}$, then we can find a sequence $\{\psi_n\}$ in $\mathcal{S}$ such that $\mu_1(\psi_n,\psi_n) = 1$ for all n but $\mu_2(\psi_n,\psi_n) \to 0$. Then, on $\mathcal{F}_2$, the sequence of operators $\{\exp[i\hat{\Omega}_2(\psi_n,\cdot)] - I_2\}$ converges strongly to zero, i.e., when applied to any fixed vector In $\mathcal{F}_2$, the resulting sequence of vectors converges to zero. However, on $\mathcal{F}_1$ the sequence of vectors $\{\exp[i\hat{\Omega}_1(\psi_n,\cdot)]\ |0>_1 - |0>_1\}$ does not converge to zero. Hence if a unitary map U satisfying eq. (4.4.1) existed, $U|0>_1 \in \mathcal{F}_2$ would contradict the strong convergence property in $\mathcal{F}_2$.

We turn our attention, now, to case (i). Equation (4.4.2) implies that μ_1 and μ_2 define "equivalent norms" on $\mathcal{S}$, i.e., a sequence in $\mathcal{S}$ is Cauchy in μ_1 if and only if it is Cauchy in μ_2. Hence, the Cauchy completions of $\mathcal{S}$ in μ_1 and μ_2 may be identified and we will denote this completion as $\mathcal{S}_\mu$ (omitting the subscript 1 or 2). The two one-particle Hilbert spaces, $\mathcal{H}_1$ and $\mathcal{H}_2$—arising from μ_1 and μ_2, respectively, by the construction described in section 3.2—thus may be viewed as different subspaces of the same (complexified, completed) space $\mathcal{S}_\mu{}^{\mathbb{C}}$.

Now, define the inner product $2\mu_1(\overline{\psi}, \chi)$ on $\mathcal{S}_\mu{}^{\mathbb{C}}$ (see eq. (3.2.18)) and let $K_1 : \mathcal{S}_\mu{}^{\mathbb{C}} \to \mathcal{H}_1$ denote the orthogonal projection map onto $\mathcal{H}_1$ arising from this inner product. Similarly, let $\overline{K}_1 : \mathcal{S}_\mu{}^{\mathbb{C}} \to \overline{\mathcal{H}}_1$ denote the orthogonal projection map onto $\overline{\mathcal{H}}_1$ in this inner product. Since $\mathcal{H}_1$ and $\overline{\mathcal{H}}_1$ are orthogonal subspaces which span $\mathcal{S}_\mu{}^{\mathbb{C}}$, we have $K_1 + \overline{K}_1 = I$ on $\mathcal{S}_\mu{}^{\mathbb{C}}$. Let $K_2 : \mathcal{S}_\mu{}^{\mathbb{C}} \to \mathcal{H}_2$ and $\overline{K}_2 : \mathcal{S}_\mu{}^{\mathbb{C}} \to \overline{\mathcal{H}}_2$ denote the similar orthogonal projection maps for the inner product $2\mu_2(\overline{\psi}, \chi)$ on $\mathcal{S}_\mu{}^{\mathbb{C}}$. By eqs. (4.2.8) and (4.4.1), unitary equivalence of the two constructions holds if and only if there exists a unitary map $U : \mathcal{F}_s(\mathcal{H}_1) \to \mathcal{F}_s(\mathcal{H}_2)$ such that for all $\psi \in \mathcal{S}$ we have

$$U[ia_1(\overline{K}_1\psi) - ia_1{}^\dagger(K_1\psi)]\,U^{-1} = ia_2(\overline{K}_2\psi) - ia_2{}^\dagger(K_2\psi) \qquad (4.4.3)$$

where a_1, $a_1{}^\dagger$ and a_2, $a_2{}^\dagger$ are the annihilation and creation operators on $\mathcal{F}_s(\mathcal{H}_1)$ and $\mathcal{F}_s(\mathcal{H}_2)$, respectively. By complex linearity, eq. (4.4.3) must hold for all $\psi \in \mathcal{S}^{\mathbb{C}}$. By continuity, eq. (4.4.3) also must hold for all $\psi \in \mathcal{S}_\mu{}^{\mathbb{C}}$. Let $A:\mathcal{H}_2 \to \mathcal{H}_1$ and $B:\mathcal{H}_2 \to \overline{\mathcal{H}}_1$ denote the restrictions of K_1 and $\overline{K}_1$, respectively, to the subspace $\mathcal{H}_2$ of $\mathcal{S}_\mu{}^{\mathbb{C}}$. Similarly, let $C:\mathcal{H}_1 \to \mathcal{H}_2$ and $D:\mathcal{H}_1 \to \overline{\mathcal{H}}_2$ denote the restrictions of K_2 and $\overline{K}_2$, respectively, to $\mathcal{H}_1$. Then, choosing $\psi \in \overline{\mathcal{H}}_1$ and writing $\chi = \overline{\psi}$, we see that unitary equivalence implies that for all $\chi \in \mathcal{H}_1$, we have

$$Ua_1(\overline{\chi})U^{-1} = a_2(\overline{C\chi}) - a_2^\dagger(\overline{D\chi}) \tag{4.4.4}$$

It follows from eq. (4.4.2) that the operators A, B, C, D are bounded. In addition, these operators satisfy a number of properties arising directly from their definition. In particular, let $\chi, \psi \in \mathcal{H}_2$. Then, we have

$$\begin{aligned}(\psi,\chi)_{\mathcal{H}_2} &= -i\Omega(\overline{\psi}, \chi) \\ &= -i\Omega(\overline{K_1\psi + \overline{K}_1\psi}, K_1\chi + \overline{K}_1\chi) \\ &= (A\psi, A\chi)_{\mathcal{H}_1} - (B\psi, B\chi)_{\overline{\mathcal{H}}_1}\end{aligned} \tag{4.4.5}$$

(No "cross terms" arise since $\Omega(\overline{\rho}, \lambda) = 0$ for all $\rho \in \overline{\mathcal{H}}_1$, $\lambda \in \mathcal{H}_1$.) Thus, we obtain

$$A^\dagger A - B^\dagger B = I \tag{4.4.6}$$

A similar calculation starting with $\chi \in \mathcal{H}_2$, $\psi \in \overline{\mathcal{H}}_2$ yields

$$A^\dagger \overline{B} = B^\dagger \overline{A} \tag{4.4.7}$$

where the bar denotes the corresponding map between the complex conjugate spaces (see Appendix A.2). In an exactly similar manner, we obtain

$$C^\dagger C - D^\dagger D = I \tag{4.4.8}$$

$$C^\dagger \overline{D} = D^\dagger \overline{C} \tag{4.4.9}$$

In addition, for $\psi \in \mathcal{H}_1$, $\chi \in \mathcal{H}_2$, we find

$$
\begin{aligned}
(\psi, A\chi)_{\mathcal{H}_1} &= -i\Omega(\overline{\psi}, K_1\chi) \\
&= -i\Omega(\overline{\psi}, K_1\chi + \overline{K}_1\chi) \\
&= -i\Omega(\overline{\psi}, \chi) \\
&= -i\Omega(\overline{K_2\psi + \overline{K}_2\psi}, \chi) \\
&= -i\Omega(\overline{K_2\psi}, \chi) \\
&= (C\psi, \chi)_{\mathcal{H}_2}
\end{aligned}
\tag{4.4.10}
$$

Thus, we have

$$A^{\dagger} = C \tag{4.4.11}$$

Similarly, we obtain

$$\overline{B}^{\dagger} = -D \tag{4.4.12}$$

Note that eqs. (4.4.6) and (4.4.8) imply that A^{-1} and C^{-1} exist as bounded operators.

A unitary map U satisfying eq. (4.4.4) with A, B, C, D satisfying eqs. (4.4.6)-(4.4.9) and (4.4.11)-(4.4.12) is known as a *Bogoliubov transformation*. We wish to determine the conditions under which a Bogoliubov transformation exists and, if it does, to obtain it explicitly. We proceed straightforwardly by supposing that U exists and attempting to solve for

$$\Psi = U\,|0\rangle_1 \tag{4.4.13}$$

We expand $\Psi \in \mathcal{F}_s(\mathcal{H}_2)$ in terms of its "n-particle amplitudes" as

$$\Psi = c\,(1, \psi^a, \psi^{ab}, \psi^{abc}, \ldots) \tag{4.4.14}$$

(The index notation used here is explained in Appendix A.3. For convenience in defining $\psi^{a_1 \cdots a_n}$, we have implicitly assumed in writing this equation that the "vacuum entry" of Ψ is nonzero; that there is no loss of generality in this assumption can be seen easily from the analysis given below.). Let $\xi \in \mathcal{H}_2$, choose $\chi = C^{-1}\xi$, and apply eq. (4.4.4) to Ψ. The left side vanishes, so we obtain, for all $\xi \in \mathcal{H}_2$,

$$0 = [a_2(\overline{\xi}) - a_2^{\dagger}(\mathcal{C}\overline{\xi})\,]\,\Psi \tag{4.4.15}$$

where the operator $\mathcal{E}:\overline{\mathcal{H}}_2 \to \mathcal{H}_2$ is defined by

$$\mathcal{E} = \overline{D}\,\overline{C}^{-1} \tag{4.4.16}$$

It follows from eq. (4.4.9) that $\mathcal{E}$ is symmetric, i.e., $\overline{\mathcal{E}}^\dagger = \mathcal{E}$.

Equation (4.4.15) yields an infinite sequence of equations for the n-particle amplitudes, $\psi^a, \psi^{ab}, \psi^{abc}, \ldots$, of Ψ. The first four of these equations are,

$$\overline{\xi}_a \psi^a = 0 \tag{4.4.17}$$

$$\sqrt{2}\,\overline{\xi}_a \psi^{ab} = (\mathcal{E}\overline{\xi})^b \tag{4.4.18}$$

$$\sqrt{3}\,\overline{\xi}_a \psi^{abc} = \sqrt{2}\,(\mathcal{E}\overline{\xi})^{(b}\psi^{c)} \tag{4.4.19}$$

$$\sqrt{4}\,\overline{\xi}_a \psi^{abcd} = \sqrt{3}\,(\mathcal{E}\overline{\xi})^{(b}\psi^{cd)} \tag{4.4.20}$$

where it is understood that these relations must hold for all $\xi \in \mathcal{H}_2$. The unique solution to eq. (4.4.17) is $\psi^a = 0$. By induction all the n-particle amplitudes vanish for odd n. On the other hand, eq. (4.4.18) implies that—when viewed as a map from $\overline{\mathcal{H}}_2 \to \mathcal{H}_2$—$\psi^{ab}$ must equal $\mathcal{E}/\sqrt{2}$. However, ψ^{ab} is not an arbitrary map: It is symmetric, $\psi^{ab} = \psi^{ba}$, and satisfies $\psi^{ab}\,\overline{\psi}_{ab} < \infty$ (see sections A.2 and A.3 of the Appendix). We already noted above that $\mathcal{E}$ is symmetric. Hence, a solution to eq. (4.4.18) exists if and only if $\mathcal{E}$ satisfies

$$\mathrm{tr}(\mathcal{E}^\dagger \mathcal{E}) < \infty. \tag{4.4.21}$$

Since both C and C^{-1} are bounded, eq. (4.4.21) is equivalent to the condition $\mathrm{tr}(D^\dagger D) < \infty$; by eq. (4.4.12) and the cyclic property of traces, it also is equivalent to $\mathrm{tr}(B^\dagger B) < \infty$. In terms of the original quantities μ_1 and μ_2, these conditions, in turn, are equivalent to the condition that the linear map $Q:\mathcal{S}_\mu \to \mathcal{S}_\mu$ defined by

$$\mu_1(\psi_1, Q\psi_2) = \mu_2(\psi_1,\psi_2) - \mu_1(\psi_1,\psi_2) \tag{4.4.22}$$

be of trace class.

If eq. (4.4.21) holds, by induction the solution obtained for Ψ is

$$\Psi = c\left(1,\, 0,\, \sqrt{\tfrac{1}{2}}\,\varepsilon^{ab},\, 0,\, \sqrt{\tfrac{3\cdot 1}{4\cdot 2}}\,\varepsilon^{(ab}\varepsilon^{cd)},\, 0,\, \ldots\right) \tag{4.4.23}$$

where ε^{ab} denotes the "two-particle state" (i.e., the element of $\mathcal{H}_2 \otimes \mathcal{H}_2$) corresponding to the map $\mathcal{E}$. It can be shown (Wald 1979a) that eq. (4.4.23) defines a (finite norm) state in $\mathcal{F}_s(\mathcal{H}_2)$. The constant c is determined (up to phase) by the condition that $\|\Psi\| = 1$, as is required for U to be unitary.

The above analysis has merely determined the action of U (if it exists) on the state $|0>_1$, given that eq.(4.4.21) holds. However, the action of U on an arbitrary one-particle state, $a_1{}^\dagger(\chi)|0>_1$, is determined by applying the adjoint of eq. (4.4.4) to $\Psi = U|0>_1$. By induction, we similarly can calculate uniquely how U must act on an arbitrary n-particle state. Since these states span $\mathcal{F}_s(\mathcal{H}_1)$ we see that U is uniquely determined. Existence and unitarity of U (given eq. (4.4.21)) then can be established straightforwardly (see Wald (1979a)). Consequently, we obtain the following theorem, which achieves the main mathematical objective set forth at the beginning of this section:

Theorem 4.4.1 : The necessary and sufficient conditions for the quantum field theories defined by the norms μ_1 and μ_2 to be unitarily equivalent are that (a) eq. (4.4.2) holds and (b) the operator Q defined by eq. (4.4.22) be of trace class (or, equivalently, that eq. (4.4.21) holds, which, in turn, is equivalent to the conditions $\mathrm{tr}(D^\dagger D) < \infty$ or $\mathrm{tr}(B^\dagger B) < \infty$).

With regard to the physical applications of our results, eq. (4.4.23) gives an explicit solution to the problem of spontaneous particle creation from the vacuum in spacetimes which are asymptotically stationary in the past and future. (Note that eq. (4.4.23) also may be interpreted as yielding the most general form of a "squeezed vacuum state"; we had promised in section 2.3 that we would provide such a formula.) One immediate consequence of eq. (4.4.23), is that particles always are created in pairs. Furthermore, since $\mathcal{E} \neq 0$ if and only if $D \neq 0$, we see that there will be a nonvanishing amplitude for spontaneous particle creation if and only if, classically, some solution which oscillates with purely positive frequency in the asymptotic past picks up a nonvanishing negative frequency part in the asymptotic future.

Note that if $\mathcal{S}$ is finite dimensional, conditions (a) and (b) of theorem 4.4.1 hold automatically, so unitary equivalence always holds. Thus, our results are consistent with the Stone-von Neumann theorem. However, if $\mathcal{S}$ is infinite dimensional these conditions need not hold, and unitarily inequivalent constructions exist. In particular, even if eq. (4.4.2) holds, eq. (4.4.21) is a nontrivial restriction. To show this—and, in particular, thereby demonstrate explicitly the existence of unitarily inequivalent constructions—let $\mu_1:\mathcal{S}\times\mathcal{S}\to\mathbb{R}$ satisfy eq. (4.2.7). (As discussed at the end of section 4.3, there is no difficulty in obtaining such a μ_1 in any globally hyperbolic, curved spacetime.) We complete $\mathcal{S}$ in this norm to obtain a real Hilbert space $\mathcal{S}_{\mu_1}$. By a Gram-Schmidt type procedure, we may choose an orthonormal basis of $\mathcal{S}_{\mu_1}$ (in the inner product $2\mu_1$) of the form $\{\psi_1, \chi_1, \psi_2, \chi_2, \ldots\}$ such that $\Omega(\psi_i, \psi_j) = \Omega(\chi_i, \chi_j) = 0$ and $\Omega(\psi_i, \chi_j) = \delta_{ij}$. We write the basis expansion of any $\psi\in\mathcal{S}_{\mu_1}$ as

$$\psi = \sum_i [\alpha_i \psi_i + \beta_i \chi_i] \tag{4.4.24}$$

Now, let $\{c_i\}$ be any bounded sequence of real numbers such that

$$\sum_i \sinh^2 c_i = \infty \tag{4.4.25}$$

Define $\mu_2:\mathcal{S}\times\mathcal{S}\to\mathbb{R}$ by

$$\mu_2(\psi, \psi') = \sum_i \{\alpha_i \alpha_i'[\cosh c_i + \sinh c_i]^2 + \beta_i \beta_i'[\cosh c_i - \sinh c_i]^2\} \tag{4.4.26}$$

for all $\psi,\psi'\in\mathcal{S}$. Then it is straightforward to verify that μ_2 satisfies eq. (4.2.7), that eq. (4.4.2) holds, but that eq. (4.4.21) fails. Thus, both μ_1 and μ_2 give rise to well defined quantum field theory constructions, but these quantum field theories are unitarily inequivalent.

As discussed at the end of section 4.2, there does not appear to be any natural choice of μ in an arbitrary globally hyperbolic curved spacetime, nor does there appear to be any natural choice of unitary equivalence class of μ's in spacetimes with a noncompact Cauchy surface. It might seem that the existence of unitarily inequivalent constructions and the lack of a criterion as to which construction to choose would pose an insurmountable obstacle to the formulation of quantum field theory in curved spacetime. However, in the next section, we shall see how the algebraic formulation quantum field theory provides a completely satisfactory means of overcoming this apparent difficulty.

4.5 The Algebraic Approach

In the approach usually taken to the formulation of a quantum theory, one first constructs states as vectors in a Hilbert space $\mathcal{F}$ (or, more generally, as density matrices on $\mathcal{F}$). One then defines observables as operators on $\mathcal{F}$ which "act upon" the states. The basic strategy of the algebraic approach is, in essence, to reverse the roles played by states and observables in the following sense: One begins by constructing observables as elements of an abstract algebra. One then defines states as objects which "act upon" observables by associating a real number to each observable. This action corresponds to "taking expectation values" in the usual approach. The crucial advantage of the algebraic approach is that it thereby allows one to treat all states—in particular, states arising in unitarily inequivalent quantum field theory constructions—on an equal footing, thereby enabling one to define the theory without the need to select a preferred construction.

The key observation which makes the algebraic approach viable for quantum field theory in curved spacetime is that—although unitarily inequivalent field theory constructions exist—the algebraic structure of the field operators in the unitarily inequivalent constructions are still "the same". In other words, even when $\{\mathcal{F}_s(\mathcal{H}_1), \hat{\Omega}_1(\psi,\cdot)\}$ and $\{\mathcal{F}_s(\mathcal{H}_2), \hat{\Omega}_2(\psi,\cdot)\}$ are unitarily inequivalent, the algebraic relations satisfied by the collection of operators $\{\hat{\Omega}_1(\psi,\cdot)\}$ on $\mathcal{F}_s(\mathcal{H}_1)$ are the same as those satisfied by the operators $\{\hat{\Omega}_2(\psi,\cdot)\}$ on $\mathcal{F}_s(\mathcal{H}_2)$. This assertion can be made mathematically precise as follows:

Let $(\mathcal{S}, \Omega)$ be a symplectic vector space with inner product $\mu:\mathcal{S}\times\mathcal{S}\to\mathbb{R}$ satisfying eq. (4.2.7). Perform the quantum field construction described in section 4.2 to obtain the Hilbert space $\mathcal{F}=\mathcal{F}_s(\mathcal{H})$ and self-adjoint operators $\hat{\Omega}(\psi, \cdot)$. Define the unitary operators $\hat{W}(\psi)$ by

$$\hat{W}(\psi) = \exp[i\hat{\Omega}(\psi,\cdot)] \tag{4.5.1}$$

Then the $\hat{W}(\psi)$ satisfy the Weyl relations:

$$\hat{W}(\psi_1)\,\hat{W}(\psi_2) = \exp[i\Omega(\psi_1,\psi_2)/2]\,\hat{W}(\psi_1+\psi_2) \tag{4.5.2}$$

$$\hat{W}^{\dagger}(\psi) = \hat{W}(-\psi) \tag{4.5.3}$$

(see eqs. (2.2.6) and (2.2.7)). Now, the collection, $\mathfrak{L}(\mathfrak{F})$, of all bounded linear maps on $\mathfrak{F}$ has the natural structure of a C*-algebra (defined in Appendix A.1), with the "*-operation" corresponding to taking adjoints. The linear maps which can be expressed as finite, complex linear combinations of $\hat{W}(\psi)$'s comprise a subalgebra of $\mathfrak{L}(\mathfrak{F})$. By taking the closure of this subalgebra in the norm provided by $\mathfrak{L}(\mathfrak{F})$, we obtain a C*-subalgebra, $\mathcal{A}$, of $\mathfrak{L}(\mathfrak{F})$, known as the *Weyl algebra.* The key fact (Slawny 1972) about this construction is that $\mathcal{A}$—viewed as an abstract C*-algebra—does not depend upon the choice of μ. In other words, although two inner products μ_1 and μ_2 satisfying eq. (4.2.7) may define unitarily inequivalent quantum field theory constructions, the C*-algebras, $\mathcal{A}_1$ and $\mathcal{A}_2$, to which they give rise are isomorphic.

This fact allows us to define the *fundamental observables* for quantum field theory in a curved spacetime to be elements of the Weyl algebra, $\mathcal{A}$, constructed from the symplectic vector space of classical solutions in the manner described above (choosing any μ which satisfies eq. (4.2.7)). We then define a *state,* ω, of the quantum field to be a linear map $\omega: \mathcal{A} \to \mathbb{C}$ satisfying the positivity condition

$$\omega(A^*A) \geq 0 \qquad (4.5.4)$$

for all $A \in \mathcal{A}$, as well as the normalization condition

$$\omega(I) = 1 \qquad (4.5.5)$$

where I denotes the identity element of $\mathcal{A}$.

Several points concerning these definitions should be noted. First, although the properties of the Weyl algebra, $\mathcal{A}$, related to its C*-algebra norm are useful for some purposes, for most applications only the *-algebra structure is needed. Natural alternative choices for a *-algebra of fundamental observables can be made (such as the algebra $\mathcal{A}'$ described in Kay and Wald (1991)). Such alternative choices give rise to technically inequivalent notions of states.

Second, the two-point function of the quantum field in state ω may be defined in terms of its action on Weyl algebra elements $W(\psi)$ by

$$<\hat{\Omega}(\psi_1,\cdot)\ \hat{\Omega}(\psi_2,\cdot)>_\omega = -\frac{\partial^2}{\partial s \partial t}\{\omega[W(s\psi_1+t\psi_2)]\exp(ist\Omega(\psi_1,\psi_2)/2)\}|_{s=t=0} \tag{4.5.6}$$

provided that the derivatives appearing on the right side of eq. (4.5.6) exist. (The two-point function is undefined if these derivatives do not exist.) Using the correspondence between solutions and test functions given by lemma 3.2.1, we may view the two-point function as a bi-distribution on spacetime, which we shall denote as $<\phi(x_1)\phi(x_2)>_\omega$. The higher n-point functions of the quantum field are defined similarly. The specification of a state, ω, on $\mathcal{A}$ corresponds, in essence, to the specification of all the smeared n-point functions of the quantum field.

Third, although $\mathcal{A}$ thus encompasses a wide class of observables, it should be noted that there are other physically relevant observables in the theory besides these. In particular, the stress-energy tensor of the quantum field is not in any sense represented as an element of $\mathcal{A}$. Thus, one should not view $\mathcal{A}$ as encompassing *all* observables of the theory; rather, one should view $\mathcal{A}$ as encompassing a "minimal" collection of observables, which is sufficiently large to enable the theory to be formulated. One may later wish to enlarge $\mathcal{A}$ and/or further restrict the notion of "state" in order to accommodate the existence of additional observables. Indeed, as we shall see in section 4.6, the restriction to states which satisfy the "Hadamard condition" appears necessary in order to define the expected stress-energy tensor.

Finally, it is worth commenting that the simple form of the positivity condition (4.5.4) on states is quite deceptive. Typically, it is highly nontrivial to check that a plausible candidate for a state actually satisfies eq. (4.5.4). In particular, since eq. (4.5.4) is nonlinear in A, it does *not* suffice to check that eq. (4.5.4) holds for a basis of $\mathcal{A}$.

What is the relationship between the algebraic notion of states defined above and the notion of states arising from the quantum field theory constructions of section 4.2? A quantum field theory construction gives rise to the Hilbert space $\mathfrak{F}_s(\mathcal{H})$ and a representation of the Weyl algebra, $\mathcal{A}$, on $\mathfrak{F}$. However, given any density matrix, $\rho:\mathfrak{F}\to\mathfrak{F}$ on any Hilbert space, $\mathfrak{F}$, which carries a representation, $\pi:\mathcal{A}\to\mathfrak{L}(\mathfrak{F})$, of $\mathcal{A}$, we obtain an algebraic state $\omega:\mathcal{A}\to\mathbb{C}$ via

$$\omega(A) = \mathrm{tr}[\rho\pi(A)] \tag{4.5.7}$$

Thus, in particular, all states arising in all quantum field theory constructions of section 4.2 give rise to algebraic states. A converse of this result is contained in the following theorem:

Theorem 4.5.1 (GNS construction): Let $\mathcal{A}$ be a C*-algebra with identity element and let $\omega:\mathcal{A}\to\mathbb{C}$ be a state. Then there exists a Hilbert space $\mathfrak{F}$, a representation $\pi:\mathcal{A}\to\mathfrak{L}(\mathfrak{F})$, and a vector $|\Psi\rangle\in\mathfrak{F}$ such that

$$\omega(A) = \langle\Psi|\,\pi(A)\,|\Psi\rangle \tag{4.5.8}$$

satisfying the additional property that $|\Psi\rangle$ is *cyclic,* i.e., the vectors $\{\pi(A)|\Psi\rangle\}$ for all $A\in\mathcal{A}$ comprise a dense subspace of $\mathfrak{F}$. Furthermore, the triple $(\mathfrak{F}, \pi, |\Psi\rangle)$ is uniquely determined (up to unitary equivalence) by these properties.

Sketch of the Construction of $(\mathfrak{F}, \pi, |\Psi\rangle)$: We use the state ω to define the bilinear map,

$$\langle A,B\rangle_{\mathcal{A}} \equiv \omega(A^* B) \tag{4.5.9}$$

taking $\bar{\mathcal{A}}\times\mathcal{A}\to\mathbb{C}$. This map is non-negative by eq. (4.5.4), so eq. (4.5.9) will define an inner product on $\mathcal{A}$ unless it happens that $\omega(A^*A) = 0$ for some $A \neq 0$. However, if that occurs, we can "factor out" by such A's, in which case eq. (4.5.9) defines a positive-definite inner product on the factor space. We then complete $\mathcal{A}$ (or the factor space) in the norm defined by (4.5.9) to get a Hilbert space $\mathfrak{F}$. We then get a representation, $\pi:\mathcal{A}\to\mathfrak{L}(\mathfrak{F})$, by having $\mathcal{A}$ act upon itself (or its factor space) by multiplication and extending this action to $\mathfrak{F}$ by continuity. Finally, the desired cyclic vector $|\Psi\rangle\in\mathfrak{F}$ simply corresponds to the identity element of $\mathcal{A}$. Further details of the proof of theorem 4.5.1 can be found, e.g., in Simon (1972). □

Given an algebraic state, ω, we refer to the collection of all density matrices on its GNS Hilbert space as the *folium* of ω.

To illustrate further the algebraic viewpoint on states and the nature of the GNS construction, let $(\mathcal{S}, \Omega)$ denote the symplectic vector space of solutions to the Klein-Gordon equation on a globally hyperbolic spacetime, and let $\mu:\mathcal{S}\times\mathcal{S}\to\mathbb{R}$ be an arbitrary (real) inner product on $\mathcal{S}$. Define $\omega:\mathcal{A}\to\mathbb{C}$ by the condition that

$$\omega[W(\psi)] = \exp(-\mu(\psi,\psi)/2) \tag{4.5.10}$$

for all $\psi\in\mathcal{S}$; ω then uniquely extends to $\mathcal{A}$ by linearity and continuity. Although ω clearly satisfies the positivity condition (4.5.4) for the elements of the form $W(\psi)$, it does not automatically satisfy this condition for (complex) linear combinations of these elements. However, it turns out that the condition

$$\mu(\psi_1,\psi_1)\,\mu(\psi_2,\psi_2) \geq \frac{1}{4}\,[\Omega(\psi_1,\psi_2)]^2 \tag{4.5.11}$$

is necessary and sufficient for the positivity of ω on $\mathcal{A}$. In particular, when μ satisfies eq. (4.2.7), ω is a state. The GNS construction for ω in this case is equivalent to the construction given in section 4.2 of the Fock space, $\mathcal{F}_s(\mathcal{H})$, with the field operators represented by eq. (4.2.8), and with $|\Psi\rangle = |0\rangle$ serving as the cyclic vector. Thus, all states in all quantum field theory constructions of the type described in section 4.2 are encompassed by the folia of states of the form (4.5.10) with μ satisfying (4.2.7).

The map $\omega:\mathcal{A}\to\mathbb{C}$ determined by eq. (4.5.10) also defines a state when μ merely satisfies eq. (4.5.11) without "saturating" this inequality, i.e., when eq. (4.2.7) fails to hold. Any state ω of the form (4.5.10) with μ satisfying (4.5.11) is referred to as a *quasi-free* state. The GNS construction for an arbitrary quasi-free state again possesses a natural Fock space structure (see, e.g., Kay and Wald (1991) for details). However, if μ fails to satisfy eq. (4.2.7), then the quasi-free state fails to be a pure state (see below) and the representation of $\mathcal{A}$ provided by the GNS construction is reducible. In stationary spacetimes, thermal equilibrium states at finite temperature are represented by quasi-free states which fail to satisfy eq. (4.2.7). In a spacetime with a noncompact Cauchy surface, states in the folium of a thermal equilibrium state do not, in general, lie in the folium of any quasi-free state with μ satisfying eq. (4.2.7).

Even the folia of all quasi-free states do not by any means encompass all algebraic states. Thus, the algebraic notion of states corresponds to an enlargement of the notion of states that goes significantly beyond that obtained by all possible quantum field theory constructions of the type considered in section 4.2. However, in my view, little, if any, harm is done by this enlargement. In particular, one can always restrict attention to any particular subclass of states which one wishes to consider. The key point is that the algebraic approach gives a (mathematically precise and elegant) means of formulating quantum field theory in curved spacetime without forcing one to make arbitrary choices, such as the specification of a particular μ satisfying eq. (4.2.7). In particular, the algebraic approach allows one to treat states arising in unitarily inequivalent constructions in a unified manner.

The above discussion has laid out the basic mathematical framework of the algebraic formulation of quantum field theory in curved spacetime. However, an issue concerning how to make physical predictions from the theory for the outcomes of measurements remains to be addressed. In the usual Hilbert space approach, an observable is represented by a self-adjoint operator, A, which, by the spectral theorem, possesses an associated family of projection operators. If the system is in the (normalized) state Ψ, the probability that a measurement of A will yield a value in the interval $I \subset \mathbb{R}$ is given by $\|P\Psi\|^2$, where P denotes the projection operator of A for the interval I. The information about this measurement can then be taken into account by "state vector reduction", i.e., by replacing Ψ with $P\Psi/\|P\Psi\|$ in subsequent evolution when a "yes-no" measurement of whether A has its value in I yields a "yes" result. This prescription leads to the following more general rule: In the Heisenberg representation, for a state represented by the density matrix ρ (normalized so that $\mathrm{tr}\rho=1$), the probability, p, that a measurement of the self-adjoint observables $A_1,\ldots,A_n$ made at the sequence of times $t_1<\ldots< t_n$ will yield results lying in the intervals $I_1,\ldots,I_n$ of $\mathbb{R}$, respectively, is given by

$$p = \mathrm{tr}\{P_n \cdots P_1 \rho \, P_1 \cdots P_n\} \tag{4.5.12}$$

where $P_1,\ldots,P_n$ denote, respectively, the projection operators of $A_1(t_1),\ldots,A_n(t_n)$ for the intervals $I_1,\ldots,I_n$. Equation (4.5.12) contains all the information available in standard quantum mechanical measurement theory. Whether or not one endorses the language of the Copenhagen interpretation used above, eq. (4.5.12) is an essential feature of standard quantum theory, and a formulation of quantum theory which fails to provide some analog of eq. (4.5.12) would be incomplete.

In the algebraic approach, the expectation value of any self-adjoint observable, $A \in \mathcal{A}$, in any state, ω, is, of course, well defined; it is given simply by $\omega(A)$. Equation (4.5.12) has precisely the form of an expectation value in the state ρ. However, this equation cannot be taken over to the algebraic approach as it stands because the projection operators, $P_1, \ldots, P_n$, need not correspond to elements of the algebra, $\mathcal{A}$. We can, of course, consider a representation, π, of $\mathcal{A}$, and obtain the projection operator, P, for the interval I of the operator representative, $\pi(A)$, of any self-adjoint element $A \in \mathcal{A}$. However, P need not be expressible as the norm limit of any sequence of operator representatives of elements of $\mathcal{A}$. (Note, however, that, as we shall discuss further below, P always can be expressed as the strong limit of a sequence of representatives of elements of $\mathcal{A}$.) Since a C*-algebra need only be closed under norm limits, there is no reason why P need be the representative of any element of $\mathcal{A}$.

One possible route toward taking over measurement theory to the algebraic approach via eq.(4.5.12) would be to impose additional conditions on $\mathcal{A}$ which would ensure that it contains elements corresponding to the projection operators associated with all of its self-adjoint elements. In particular, one could require that $\mathcal{A}$ have the structure of a von Neumann algebra. (A *von Neumann algebra* may be defined as a C*-subalgebra of the bounded linear operators on a Hilbert space which contains the identity and is closed in the weak topology on operators; see, e.g., Lansford (1971) for considerable further discussion.) However, for quantum field theory in curved spacetime, such an additional requirement on $\mathcal{A}$ would constitute a significant step backwards towards the Hilbert space approach, since it would raise the issue of which von Neumann (or other) algebra to choose, which would be similar in nature to the issue within the Hilbert space approach of which representation of the Weyl algebra to choose.

In fact, however, measurement theory can be formulated without difficulty in the algebraic approach assuming only C*-algebra structure for $\mathcal{A}$. For an arbitrary state ω (normalized so that $\omega(I) = 1$), the probability, p, that a measurement of the self-adjoint observables $A_1, \ldots, A_n$ made at the sequence of times $t_1 < \ldots < t_n$ will yield results lying in the intervals $I_1, \ldots, I_n$ of $\mathbb{R}$ can be defined by the following formula:

$$p = \lim_{i_1 \cdots i_n \to \infty} \omega[(Q_1)_{i_1} \cdots (Q_n)_{i_n} (Q_n)_{i_n} \cdots (Q_1)_{i_1}] \,. \qquad (4.5.13)$$

Here, for each fixed $k = 1,\ldots,n$, $\{(Q_k)_{i_k}\}$ (with $i_k = 1,2,\ldots$) is any sequence of polynomials in A_k such that the polynomial functions $\{(Q_k)_{i_k}(x)\}$ are uniformly bounded on the spectrum of A_k and converge pointwise on the spectrum of A_k to the characteristic function of the interval I_k. To show that eq. (4.5.13) is a sensible expression, we must show that the limit on the right side of this equation exists, is independent of the choice of $\{(Q_k)_{i_k}\}$, and that the resulting probabilities, p, satisfy the necessary positivity and sum rule properties. To do so, we evaluate the right side of eq. (4.5.13) in the GNS representation. We have

$$\begin{aligned}
&\omega[(Q_1)_{i_1}\cdots(Q_n)_{i_n}(Q_n)_{i_n}\cdots(Q_1)_{i_1}\cdot] \\
&\qquad = \langle\Psi|\pi[(Q_1)_{i_1}]\cdots\pi[(Q_n)_{i_n}]\,\pi[(Q_n)_{i_n}\cdots\pi[(Q_1)_{i_1}]|\Psi\rangle \\
&\qquad = \langle\Psi|(Q_1)_{i_1}[\pi(A_1)]\cdots(Q_n)_{i_n}[\pi(A_n)]\,(Q_n)_{i_n}[\pi(A_n)]\cdots(Q_1)_{i_1}[\pi(A_1)]|\Psi\rangle \\
&\qquad = \mathrm{tr}\{(Q_n)_{i_n}[\pi(A_n)]\cdots(Q_1)_{i_1}[\pi(A_1)]\,\rho\,(Q_1)_{i_1}[\pi(A_1)]\cdots(Q_n)_{i_n}[\pi(A_n)]\}
\end{aligned} \tag{4.5.14}$$

where $|\Psi\rangle$ is the GNS representative of ω, $(Q_k)_{i_k}[\pi(A_k)]$ denotes the polynomial function of $\pi(A_k)$ corresponding to $(Q_k)_{i_k}$, and, in the last line, $\rho = |\Psi\rangle\langle\Psi|$. It follows directly from the functional calculus form of the spectral theorem (see theorem VII.2 of Reed and Simon (1980)) that each $(Q_k)_{i_k}[\pi(A_k)]$ converges strongly to the projection operator P_k for $\pi(A_k)$ associated with the interval I_k. This ensures that the limit of the right side of eq. (4.5.13) exists and equals what would be obtained from eq.(4.5.12) in the GNS representation—or, indeed any other representation of $\mathcal{A}$ in which the state ω can be realized by a density matrix ρ. In other words, eq. (4.5.13) is equivalent to the rule that probabilities, p, be computed by passing to the GNS (or any other) representation and using the standard Hilbert space rules. (The representation-independent form of eq.(4.5.13) ensures that the value of p is independent of the choice of representation.) Thus, the probabilities for outcomes of any sequence of measurements of observables in $\mathcal{A}$ are well defined in the algebraic approach, assuming only C*-algebra structure for $\mathcal{A}$.

Note that eq. (4.5.13) can be viewed as providing a rule for how to "reduce" an algebraic state ω following a measurement. If "yes" answers are obtained for the measurement of whether the observable A lies in the interval I, the new algebraic state ω' which encodes this information is given by,

$$\omega'(B) = \lim_{i\to\infty} \omega[Q_i(A)\ B\ Q_i(A)]/p \qquad (4.5.15)$$

for all $B \in \mathcal{A}$, where $Q_i(A)$ is any "polynomial approximate" of the projection for the interval I (i.e., $\{Q_i(x)\}$ is uniformly bounded on the spectrum of A and converges pointwise on the spectrum of A to the characteristic function of the interval I), and p is given by eq. (4.5.13) (which, in this case, reduces simply to the limit of $\omega[Q_i(A)]$).

As noted above, the algebraic framework presented above gives rise to a significant enlargement of the class of states admitted by the theory. It might be thought that this enlargement could yield to states with very different physical properties with respect to observables in $\mathcal{A}$ than any of the states arising from the quantum field theory constructions of section 4.2. Indeed, it might even be thought that two unitarily inequivalent quantum field theory constructions of section 4.2 (arising from different choices of μ) could give rise to states with very different properties from each other with respect to observables in $\mathcal{A}$. However, it turns out that this is not the case. This conclusion is expressed by the following theorem, which follows from results in Fell (1960) together with the fact that the Weyl algebra is simple (see, e.g. Simon (1972)) and hence all of its representations are faithful.

Theorem 4.5.2: Let $(\mathcal{F}_1, \pi_1)$ and $(\mathcal{F}_2, \pi_2)$ be (possibly unitarily inequivalent) representations of the Weyl algebra $\mathcal{A}$. Let $A_1,\ldots,A_n \in \mathcal{A}$ and let $\varepsilon_1,\ldots,\varepsilon_n > 0$. Let ω_1 be an algebraic state corresponding to a density matrix on $\mathcal{F}_1$. Then there exists a state ω_2 corresponding to a density matrix on $\mathcal{F}_2$ such that for all i = 1,...,n, we have

$$|\omega_1(A_i) - \omega_2(A_i)| < \varepsilon_i \qquad (4.5.16)$$

Theorem 4.5.2 directly shows that, although two representations of $\mathcal{A}$ may be unitarily inequivalent, the determination of a finite number of expectation values of observables in $\mathcal{A}$—each made with finite accuracy—cannot distinguish between different representations. Now, as discussed above, $\mathcal{A}$ need not contain elements corresponding to projection operators associated with $A_1,\ldots,A_n$, so theorem 4.5.2 does not imply that a state ω_2 corresponding to a density matrix on $\mathfrak{F}_2$ can be found such that the probabilities for a given, finite sequence of measurements in state ω_2 (see eq. (4.5.13)) are arbitrarily close to the corresponding probabilities for ω_1. However, the notion of measurement implicit in eq. (4.5.13) is highly idealized in that it assumes that one can distinguish with infinite precision whether or not the observable A lies in the interval I. A much more realistic model of a physically achievable measurement in the usual Hilbert space formulation would correspond to replacing each projection operator, P_k, in eq. (4.5.12) with some continuous function, $f_k(A_k)$, of the observable A_k, where $f_k:\mathbb{R}\rightarrow\mathbb{R}$ agrees with the characteristic function of the interval I_k except in an (arbitrarily small) neighborhood of the endpoints of I_k. However, if f_k is continuous, then $f_k(A_k)$ can be expressed as the norm limit of polynomials in A_k (see, e.g., theorem VII.1 of Reed and Simon (1980)). This implies that in the algebraic framework, $f_k(A_k)$ will be represented as an element of $\mathcal{A}$, and hence formula (4.5.12) with P_k replaced by $f_k(A_k)$ can be taken over to the algebraic framework without modification. Theorem 4.5.2 then implies that if these measurements are made to any finite accuracy, it is impossible to determine the representation to which the state belongs.

Thus, if observables in $\mathcal{A}$ were the only measurable quantities in quantum field theory, theorem 4.5.2 could be interpreted as saying that—given the limitation of finitely many physically realistic measurements with finite accuracy—the different representations of $\mathcal{A}$ are "physically equivalent" (Haag and Kastler 1964), and the choice of representation of $\mathcal{A}$ is physically irrelevant. In that case, it could be argued that we might as well choose any μ (arbitrarily) and define quantum field theory in curved spacetime via the construction of section 4.2 using that μ, since no realistic experiment can ever tell us that the quantum field is not in a density matrix state arising from that particular construction. However, in addition to the mathematical inelegance of proceeding in this manner, this viewpoint is not viable because there are additional observables which one wishes to define in the theory—such as the stress-energy tensor—which are not represented in $\mathcal{A}$, and two representations need not be "physically equivalent" with respect to these additional observables. Indeed, states which fail to satisfy the Hadamard condition and thus are "too pathological" to allow their expected stress-energy to be defined (see section 4.6 below) can be found arbitrarily close in the sense of theorem 4.5.2 to states which satisfy the Hadamard condition. However, if these additional observables could be incorporated into an enlarged algebra in such a way that theorem 4.5.2 continued to hold, then the choice of representation (of this enlarged algebra) would be, in essence, physically irrelevant.

In the light of theorem 4.5.2, we now may give a sensible interpretation of the situation where the S-matrix, U, relating two quantum field constructions, fails to exist. If eqs. (4.4.2) and (4.4.21) hold—and, hence, the S-matrix exists—then $U|0\rangle_1$ is given explicitly by eq. (4.4.23), and the action of U on the other vectors in $\mathfrak{F}_1$ is given by similar expressions. However, when the S-matrix fails to exist, the right side of eq. (4.4.23) does not define a normalizable state in $\mathfrak{F}_2$. Nevertheless, suitable "truncated versions" of this expression yield normalizable states in $\mathfrak{F}_2$ which approximate the state $|0\rangle_1$ (viewed as an algebraic state) to arbitrarily good accuracy in the sense of theorem 4.5.2. (To obtain such a "truncated version" of eq. (4.4.23), we first find a finite dimensional subspace, $\mathcal{S}'$, of $\mathcal{S}$ such that $A_1,\ldots, A_n$ are well approximated by sums of elements of the form (4.5.1) with $\psi \in \mathcal{S}'$. Then, on the right side of eq. (4.4.23), we replace ε^{ab} by ε'^{ab}, where ε'^{ab} denotes the projection of ε^{ab} onto $K_2\mathcal{S}' \otimes K_2\mathcal{S}'$.) Thus, the formal computation of U in situations where eq. (4.4.2) and/or eq. (4.4.21) do not hold is far from a fruitless exercise: It can be used to determine states in $\mathfrak{F}_2$ which approximate a given state in $\mathfrak{F}_1$. A similar "truncation" in quantum electrodynamics of the formal expression for the S-matrix can be used to sensibly interpret the "infra-red catastrophe."

We conclude this section by introducing the notions of "pure" and "mixed" states in the algebraic framework, the notion of a subalgebra associated with a spacetime region, and the notion of the restriction of a state to such a region. An algebraic state, ω, is said to be *mixed* if it can be expressed in the form

$$\omega = c_1\, \omega_1 + c_2\, \omega_2 \tag{4.5.17}$$

where $c_1, c_2 > 0$ and ω_1 and ω_2 are states (with $\omega_1 \neq \omega_2$); otherwise ω is said to be *pure*. Note that the GNS construction always expresses any state (pure or mixed) as a vector state (as opposed to density matrix) in a Hilbert space carrying a representation of $\mathcal{A}$. However, the distinction between pure and mixed states does enter the GNS construction in that the representation of $\mathcal{A}$ will be irreducible if and only if the state is pure (see, e.g., Simon (1972)). A quasifree state (defined above by eq. (4.5.10) with μ satisfying eq. (4.5.11)) will be pure if and only if μ satisfies eq. (4.2.7) (see Kay and Wald (1991)). Thus, in the algebraic language, our original quantum field constructions of section 4.2 correspond precisely to considering states which lie in the folia of pure, quasi-free states.

Let (M, g_{ab}) be a globally hyperbolic spacetime and let $\mathcal{O} \subset M$ be any open subset. The space, $\mathcal{T}_{\mathcal{O}}$, of test functions with support contained in $\mathcal{O}$ then is naturally a subspace of the space, $\mathcal{T}$, of all test functions in M. Using the relation between test functions and solutions given by lemma 3.2.1, we thereby can associate with $\mathcal{O}$ a subspace, $\mathcal{S}_{\mathcal{O}}$, of $\mathcal{S}$. By taking the norm closure of the span of the elements of the form (4.5.1) with ψ in $\mathcal{S}_{\mathcal{O}}$, we obtain, in turn, a C*-subalgebra, $\mathcal{A}_{\mathcal{O}} \subset \mathcal{A}$, associated with $\mathcal{O}$. Note that if $(\mathcal{O}, g_{ab})$—viewed as a spacetime in its own right—is globally hyperbolic with a Cauchy surface of the form $\mathcal{O} \cap \Sigma$, where Σ is a Cauchy surface for M, then $\mathcal{A}_{\mathcal{O}}$ is just the Weyl algebra of the spacetime $(\mathcal{O}, g_{ab})$.

The above construction yields a net of C*-algebras, $\{\mathcal{A}_{\mathcal{O}}\}$, indexed by the spacetime regions $\mathcal{O}$ (Dimock 1980). It is easily seen that if $\mathcal{O} \subset \mathcal{O}'$, then $\mathcal{A}_{\mathcal{O}}$ can be viewed as a subalgebra of $\mathcal{A}_{\mathcal{O}'}$ in a natural manner. On the other hand, if no point of $\mathcal{O}$ can be connected to a point of $\mathcal{O}'$ by a causal curve, then by the generalization of eq. (3.2.30) to curved spacetime (see section 4.2), we find that $\mathcal{A}_{\mathcal{O}}$ commutes with $\mathcal{A}_{\mathcal{O}'}$. This basic structure of $\mathcal{A}$ as a net, $\{\mathcal{A}_{\mathcal{O}}\}$, of local C*-subalgebras plays an important role in the general algebraic formulation of quantum field theory; see Haag (1992) for considerable further discussion.

Let ω be a state on the Weyl algebra, $\mathcal{A}$, of (M, g_{ab}), and let $\mathcal{O} \subset M$ be open. Then the restriction of ω to $\mathcal{A}_{\mathcal{O}}$ gives rise to a state $\omega_{\mathcal{O}}$ on the algebra $\mathcal{A}_{\mathcal{O}}$. We define the *domain of determinacy*, $\mathcal{D}(\mathcal{O})$, of $\mathcal{O}$ by

$$\mathcal{D}(\mathcal{O}) = \left\{x \in M \mid \psi(x)=0 \text{ for all } \psi \in \mathcal{S} \text{ such that } \psi=0 \text{ in } \mathcal{O}\right\} \quad (4.5.18)$$

Then, clearly, if $\mathcal{D}(\mathcal{O}) \neq M$, we have $\mathcal{S}_{\mathcal{O}} \neq \mathcal{S}$; indeed, even when $\mathcal{D}(\mathcal{O}) = M$, we may have $\mathcal{S}_{\mathcal{O}} \neq \mathcal{S}$. If $\mathcal{S}_{\mathcal{O}} \neq \mathcal{S}$, it does not automatically follow that $\mathcal{A}_{\mathcal{O}} \neq \mathcal{A}$, since the construction of these algebras involves the taking of a norm closure. However, it is easily seen that $\mathcal{A}_{\mathcal{O}} \neq \mathcal{A}$ whenever there exists an nonempty open set $\mathcal{O}'$ such that no point of $\mathcal{O}$ can be connected to a point of $\mathcal{O}'$ by a causal curve.

If $\mathcal{A}_{\mathcal{O}} \neq \mathcal{A}$, then, in general, $\omega_{\mathcal{O}}$ will be a mixed state even if ω

was pure. Indeed, in the case of Minkowski spacetime, suppose that $\mathcal{O}$ is any open region for which there exists a nonempty open region, $\mathcal{O}'$, all of whose events are spacelike related to the events $\mathcal{O}$. Then the Reeh-Schlieder theorem (see, e.g., Streater and Wightman (1964)) implies that the restriction of the ordinary vacuum state, $|0\rangle$, to $\mathcal{O}$ defines a mixed state. The Unruh effect—to be considered in the next chapter—provides an excellent illustration of this phenomenon.

4.6 The Stress-Energy Tensor

In the previous section, we saw how the algebraic approach allows one to formulate the quantum theory of a scalar field, despite the existence of unitarily inequivalent representations and the nonexistence of any clear criteria for picking out a preferred representation. The theory thereby obtained directly describes the expectation values of all observables appearing in the Weyl algebra, $\mathcal{A}$. (These expectation values are, in essence, equivalent to the specification of n-point correlation functions of the quantum field [see eq. (4.5.6)].) As we saw in the previous section, the rules of measurement theory for observables in $\mathcal{A}$ also can be formulated without difficulty. In addition, the Hamiltonian (up to a multiple of the identity) also is automatically represented in the theory in the following manner: Classical time evolution (generated by the classical Hamiltonian) gives rise to a linear mapping $\tau_t:\mathcal{S}\rightarrow\mathcal{S}$, which preserves the symplectic structure Ω. This symplectic map on $\mathcal{S}$ gives rise, in turn, to an automorphism $\alpha_t:\mathcal{A}\rightarrow\mathcal{A}$ of the Weyl algebra. This automorphism corresponds to the transformation of operator observables on a Hilbert space in the Heisenberg picture, and thus may be viewed as defining the Hamiltonian in the quantum theory up to a multiple of the identity, in the same manner as previously described in section 2.3.

However, there are many additional field observables in the classical theory which one may wish to represent in the quantum theory. Primary among these is the stress-energy tensor, which, for the classical Klein-Gordon field, is given by

$$T_{ab} = \nabla_a\phi\,\nabla_b\phi - \frac{1}{2}\,g_{ab}(\nabla_c\phi\,\nabla^c\phi + m^2\phi^2) \qquad (4.6.1)$$

This quantity is of interest in its own right since it describes the local energy, momentum, and stress properties of the field. However, it also is of direct relevance for describing the back-reaction of the quantum field on the spacetime geometry. Classically, matter influences gravity via its stress-energy tensor (appearing as a source term in Einstein's equation) and, as discussed further at the end of this section, one would expect that, semiclassically, back-reaction would be described by the "semiclassical Einstein equation",

$$G_{ab} = 8\pi \langle T_{ab} \rangle \tag{4.6.2}$$

Thus, it is of considerable interest to determine the expectation value, $\langle T_{ab} \rangle$, of the stress-energy tensor in physically relevant states.

Equation (4.6.1) expresses T_{ab} as a simple, quadratic expression in the field ϕ. If the quantum field $\hat{\phi}$ were well defined as an operator at each $x \in M$ (or, equivalently, if the Weyl algebra contained representatives of $\hat{\phi}(x)$ rather than smeared field operators) it would be entirely straightforward to define $\hat{T}_{ab}$ in the quantum theory. However, since $\hat{\phi}$ is well defined only as a distribution on spacetime, eq. (4.6.1) involves taking the product of two distributions at the same spacetime point, which is intrinsically ill-defined mathematically. Consequently, some sort of "regularization procedure" is needed to define the stress-energy tensor in the quantum theory.

One approach toward doing this would be to attempt to enlarge the algebra of observables to include representatives of (suitably "smeared") stress-energy tensors. Indeed, it probably would be necessary to proceed in this manner in order to extend the rules of measurement theory described in the previous section to the stress-energy observables. However, I am not aware of any significant progress in this direction. Alternatively, one could simply seek to define the expected stress-energy tensor for a suitable class of states, ω, on the original Weyl algebra, $\mathcal{A}$, without attempting to put any algebraic structure on the stress-energy observables. Although this approach need not yield any well defined rules for determining the probabilities for the results of measurements of the stress-energy tensor, it automatically provides a rule for calculating the quantity which appears on the right side of the semiclassical back-reaction equation (4.6.2). We shall follow this route below and shall denote the expected stress-energy in (algebraic) state ω as $\langle T_{ab} \rangle_\omega$

or, more simply, as $<T_{ab}>$. Some restrictions should be expected on the class of states for which $<T_{ab}>$ can be defined, since one would expect $<T_{ab}>$ to be singular in any state with bad "ultraviolet behavior". We shall see that the Hadamard condition (defined below) provides a restriction of exactly this sort on states.

The difficulties with defining $<T_{ab}>$ are present already in the standard quantum field theory construction in Minkowski spacetime. One could attempt to calculate $<T_{ab}>$ by substituting the formal mode sum expression (3.1.19) for $\hat{\phi}(x)$ into eq. (4.6.1). When one does so, one gets an infinite contribution arising from the sum of terms of the form $a_i a_i^\dagger$. When integrated over space, this contribution corresponds to the sum of the zero-point energies of the infinite collection of harmonic oscillators which comprise the field. This suggests a simple cure for this difficulty: normal ordering. One simply puts ("by hand") all annihilation operators to the right of creation operators in the formal expression for $<T_{ab}>$. This normal ordering prescription is designed to make $<0 \mid T_{ab} \mid 0> = 0$. Indeed, the prescription may be viewed as "subtraction of the vacuum stress-energy" from the formal expression for $<T_{ab}>$.

An alternative formulation of the normal ordering prescription can be given which makes more clear the sense in which it corresponds to "subtraction of the vacuum stress-energy". To keep the formulas simple, we first consider this prescription for the calculation of $<\phi^2>$ rather than $<T_{ab}>$. Like the stress energy tensor, the calculation of $<[\phi(x)]^2>$ involves taking the product of two distributions at the same spacetime point, and the formal mode-sum expression for it diverges. However, $<\phi(x)\phi(x')>$ makes perfectly good mathematical sense as a bi-distribution. Now, for physically reasonable states in the standard Fock space—in particular, states with finitely many particles, each of which is in a state corresponding to a smooth mode function—the singular behavior of this bi-distribution as $x' \to x$ is the same as for $<0 \mid \phi(x)\phi(x') \mid 0>$. More precisely, for such states, the difference,

$$F(x,x') \equiv <\phi(x)\phi(x')> - <0 \mid \phi(x)\phi(x') \mid 0> \qquad (4.6.3)$$

is a smooth function of x and x'. Hence, after performing this "vacuum subtraction", the coincidence limit may be taken. We then define

$$<\phi^2(x)> = \lim_{x'\to x} F(x,x') \qquad (4.6.4)$$

A similar "point-splitting" prescription can be given for defining $<T_{ab}>$, wherein one takes appropriate derivatives of $F(x,x')$ with respect to x and x' prior to taking the coincidence limit. More precisely, we define

$$<T_{ab}(x)> = \lim_{x'\to x} \left\{\nabla_a \nabla'_{b'} F(x,x') - \tfrac{1}{2} g_{ab}[(\nabla_c \nabla'^{c'} + m^2)F(x,x')]\right\} \qquad (4.6.5)$$

Here the unprimed indices refer to tensors over the tangent space at x, while the primed indices refer to tensors over the tangent space at x'. These two tangent spaces can be identified via parallel transport along the unique geodesic connecting x' and x, and this identification is understood in eq. (4.6.5) prior to taking the coincidence limit, so that the term inside the braces can be viewed as a tensor in the tangent space at x. (A similar identification of the tangent spaces also will be understood in curved spacetime when x' is sufficiently close to x that it lies within a normal neighborhood of x.) It can be readily verified that the "point-splitting" prescription in Minkowski spacetime defined by eq. (4.6.5) is equivalent to normal ordering.

Although normal ordering gives an entirely satisfactory prescription for defining $<T_{ab}>$ on a suitable class of states lying in the standard Fock representation, there is no satisfactory straightforward generalization of it to curved spacetime. Namely, as already emphasized several times in this chapter, in a general curved spacetime there is no "preferred vacuum state". Furthermore, even in cases (such as stationary spacetimes) where a natural vacuum state can be picked out, one would not expect $<T_{ab}>$ to vanish for this state, i.e., "vacuum polarization" effects would be expected to make $<T_{ab}> \neq 0$. Thus, a normal ordering prescription involving "subtracting off the vacuum energy" for some globally defined vacuum state cannot be expected to be valid in a general, globally hyperbolic curved spacetime. However, note that the above "point-splitting" prescription sensibly defines the *differences* in the expected stress energy between two states, i.e., $<T_{ab}>_1 - <T_{ab}>_2$ is well defined for any two algebraic states ω_1, ω_2 which possess a two-point function (see eq. (4.5.6)) and are such that $<\phi(x)\phi(x')>_1 - <\phi(x)\phi(x')>_2$ is a smooth function.

How is $<T_{ab}>$ to be defined? In the absence of an obvious prescription it is useful to take an axiomatic approach and seek to determine $<T_{ab}>$ by the properties one would expect it to satisfy. The following is a list of such properties:

1) The natural, well defined expression for the difference in expected stress energy between two states should be valid, i.e., whenever $<\phi(x)\phi(x')>_1 - <\phi(x)\phi(x')>_2$ is a smooth function, $<T_{ab}>_1 - <T_{ab}>_2$ should be given by the above "point splitting" prescription.

2) The expected stress-energy should be local with respect to the state of the field in the following sense: Let (M, g_{ab}) and (M', g'_{ab}) be globally hyperbolic spacetimes and suppose a globally hyperbolic, open neighborhood, $\mathcal{O}$, of $p \in M$—with Cauchy surface of the form $\mathcal{O} \cap \Sigma$, where Σ is a Cauchy surface for M—is isometric to a globally hyperbolic open neighborhood $\mathcal{O}' \subset M'$ with Cauchy surface of the form $\mathcal{O}' \cap \Sigma'$, where Σ' is a Cauchy surface for M'. Let $i: \mathcal{O} \to \mathcal{O}'$ denote this isometry and let $p' = i(p)$. Using this isometry, i, we may identify the Weyl subalgebra $\mathcal{A}_{\mathcal{O}} \subset \mathcal{A}$ for M with the Weyl subalgebra $\mathcal{A}'_{\mathcal{O}'} \subset \mathcal{A}'$ for M'. Now, suppose the states ω on $\mathcal{A}$ and ω' on $\mathcal{A}'$ are such that under this identification their restrictions to $\mathcal{A}_{\mathcal{O}}$ and $\mathcal{A}'_{\mathcal{O}'}$, respectively, are equal. Then we require that under the identification of $\mathcal{O}$ and $\mathcal{O}'$ given by i, $<T_{ab}>_\omega$ at p should equal $<T_{ab}>_{\omega'}$ at p'.

3) For all states, we have $\nabla^a <T_{ab}> = 0$.

4) In Minkowski spacetime, we have $<0|T_{ab}|0> = 0$.

These four properties correspond to the first four axioms of Wald (1977). (The fifth axiom of that reference cannot be satisfied, as was shown in Wald (1978a).) The only notable change is that the "causality axiom" of Wald (1977) has been reformulated as the above "locality property". This reformulation is much more in keeping with the spirit of the algebraic approach adopted here.

The first two of the above properties are the key ones, since they uniquely determine the expected stress-energy tensor up to the addition of local curvature terms. To see this, let $<T_{ab}>$ and $<\tilde{T}_{ab}>$ denote two prescriptions for the expected stress-energy tensor which satisfy properties (1) and (2). Then, for any two states ω_1, ω_2 in a given fixed curved spacetime, by property (1), we have

$$<T_{ab}>_1 - <T_{ab}>_2 = <\tilde{T}_{ab}>_1 - <\tilde{T}_{ab}>_2 \tag{4.6.6}$$

However, eq.(4.6.6) implies that in any fixed spacetime, (M, g_{ab}), the quantity

$$t_{ab} \equiv \langle T_{ab}\rangle_\omega - \langle \tilde{T}_{ab}\rangle_\omega \qquad (4.6.7)$$

actually is independent of the state of the field ω. Property (1) does not restrict the manner in which t_{ab} may depend upon the spacetime (M, g_{ab}). However, since t_{ab} is independent of state, property (2) immediately implies that t_{ab} at $p \in M$ depends only on the spacetime geometry in an arbitrarily small open neighborhood of p. Thus, t_{ab} is a "local curvature term", and, hence, eq. (4.6.7) shows that any two prescriptions satisfying properties (1) and (2) can differ at most by a local curvature term. Property (3) then further implies that $\nabla^a t_{ab} = 0$, and property (4) implies that $t_{ab} = 0$ in Minkowski spacetime. Thus, we obtain the following uniqueness theorem:

Theorem 4.6.1: Let $\langle T_{ab}\rangle$ and $\langle \tilde{T}_{ab}\rangle$ denote two prescriptions for the expected stress-energy tensor which satisfy properties (1)-(4) above. Then $t_{ab} \equiv \langle T_{ab}\rangle_\omega - \langle \tilde{T}_{ab}\rangle_\omega$ is a conserved local curvature term, i.e., t_{ab} is independent of the state ω, satisfies $\nabla^a t_{ab} = 0$, and its value at any event p depends only upon the spacetime geometry in an arbitrarily small neighborhood of p—with $t_{ab}(p) = 0$ if the geometry in a neighborhood of p is flat.

If we assume, in addition, that t_{ab} has dimension $(\text{length})^{-4}$, that the mass, m, of the Klein-Gordon field is the only dimensionful parameter upon which t_{ab} can depend, and that the limit $m \to 0$ is continuous, then t_{ab} must be constructed out of terms which either are linear or quadratic in the curvature. Conservation of t_{ab} then implies that the only candidate terms are $m^2 G_{ab}$ and the two independent, conserved, local curvature terms which can be obtained by functionally differentiating the quadratic curvature Lagrangians, R^2 and $R_{ab}R^{ab}$, with respect to the metric. The ambiguity in $\langle T_{ab}\rangle$ resulting from the term $m^2 G_{ab}$ can be "absorbed" in the semiclassical back-reaction equation (4.6.2) by a redefinition of Newton's constant G. However, the ambiguity resulting from the quadratic curvature terms nontrivially affects this equation (see below). Note that this ambiguity is the same as that arising at one-loop order from the nonrenormalizability of quantum gravity. As discussed further at the end of this section, this ambiguity presumably will not be resolved until a complete quantum theory of gravity is available.

Theorem 4.6.1 establishes uniqueness of $<T_{ab}>$ (up to addition of a conserved local curvature term) but says nothing concerning existence of an acceptable prescription. A "point-splitting" prescription wherein, in each spacetime, one subtracts from $<\phi(x)\phi(x')>$ its expectation value in some fixed state will automatically satisfy property (1), but will not satisfy property (2). However, a "point-splitting" prescription wherein, in each spacetime, one subtracts from $<\phi(x)\phi(x')>$ a *locally constructed* bi-distribution, $H(x,x')$, with suitable singularity structure in the coincidence limit will automatically satisfy both properties (1) and (2). In addition, if $H(x,x')$ satisfies the wave equation in both x and x' and if it equals $<0|\phi(x)\phi(x')|0>$ in Minkowski spacetime, then properties (3) and (4) also will be satisfied. Thus, existence of a suitable prescription for defining $<T_{ab}>$ will be established if a suitable bi-distribution $H(x,x')$ can be constructed. As we shall see, there is a difficulty in obtaining a locally constructed $H(x,x')$ which satisfies the wave equation in both x and x', but existence of a suitable prescription for defining $<T_{ab}>$ can be established nevertheless.

A prescription for constructing local bi-distributional solutions to elliptic and hyperbolic equations was given by Hadamard (1923) for the purpose of proving existence and uniqueness results for such equations. Hadamard's algorithm was motivated by the attempt to construct bi-distributions analogous to the standard Green's function for Laplace's equation in a n-dimensional, flat Euclidean space, which takes the form $1/\sigma^s(x,x')$, where σ denotes the squared geodesic distance between x and x' and $s = (n-2)/2$. Thus, the basic idea is to seek bi-solutions with "leading order singularity" of this form in the coincidence limit, $x \to x'$. Note that in a curved (Riemannian or Lorentzian) space, $\sigma(x,x')$ is not, in general, well defined, since it can happen that a given pair of points (x,x') may be connected by many geodesics or no geodesics at all. However, the manifold always may be covered by convex normal neighborhoods (see, e.g., Hawking and Ellis (1973)), and within each of these neighborhoods $\sigma(x,x')$ is well defined. Since, ultimately, our algorithm for defining $<T_{ab}>$ will involve taking the coincidence limit, $x \to x'$, the lack of a global definition of σ does not cause any problem.

In the case of interest to us—namely the Klein-Gordon equation in a four-dimensional, globally hyperbolic, Lorentz signature spacetime (M, g_{ab})—we must specify "$1/\sigma$" more carefully in order to define it as a distribution, since σ is singular whenever x and x' are connected by a null geodesic, not merely when they are coincident. The choice we make of $1/[\sigma+2i\varepsilon(t-t')+\varepsilon^2]$ is motivated by the form of the vacuum two-point function, $\langle 0|\phi(x)\phi(x')|0\rangle$, of the Klein-Gordon field in Minkowski spacetime. (Here t is any global time function, and it is understood that the limit as $\varepsilon \downarrow 0$ should be taken after applying this quantity to test functions.) Thus, we choose the precise form of the Hadamard anzatz in our case to be

$$H(x,x') = \frac{U(x,x')}{(2\pi)^2[\sigma+2i\varepsilon(t-t')+\varepsilon^2]} + V(x,x')\ln[\sigma+2i\varepsilon(t-t')+\varepsilon^2] + W(x,x') \tag{4.6.8}$$

with

$$V(x,x') = \sum_{j=0}^{\infty} v_j(x,x')\,\sigma^j \tag{4.6.9}$$

$$W(x,x') = \sum_{j=0}^{\infty} w_j(x,x')\,\sigma^j \tag{4.6.10}$$

where U, v_i, w_k are smooth functions of (x,x'), with U(x,x) = 1. As we shall see below, the possibility that the series in eqs. (4.6.9) and (4.6.10) need not, in general, converge is not of concern here, since only the first few terms in these series will be relevant to our algorithm for defining $\langle T_{ab}\rangle$.

To proceed, one formally substitutes H into the Klein-Gordon equation in x (at fixed x') and equates the coefficients of the explicitly appearing powers of σ and $\ln\sigma$ to zero. One thereby obtains a sequence of ordinary differential equations for U, v_i, and w_k along the geodesic connecting x and x'. The equations for U and v_i can be solved recursively and have a unique regular solution which can be expressed in terms of integrals along the geodesic connecting x and x'. (The square of the solution for U is known as the "van Vleck-Morette determinant".) Further details concerning this construction can be found in Garabedian (1964). It can be shown (Garabedian 1964) that U and V are automatically symmetric in x and x'. The function w_0 is undetermined by the system of equations, although all w_k for k>0 are uniquely determined in a local manner once w_0 has been specified. However, as explicitly seen in Wald (1978a), the w_k

need not be symmetric in x and x' even if w_0 is chosen to be symmetric. This is relevant because if W fails to be symmetric, H need not satisfy the Klein-Gordon equation in x' at fixed x. In the massive case, the "DeWitt-Schwinger" algorithm (see, e.g., DeWitt (1975) or Birrell and Davies (1982)) for constructing a short-distance asymptotic expansion for a Green's function—which is based upon an asymptotic expansion for the "heat kernel" of an elliptic operator—in effect provides a local choice of w_0 which produces w_k which are symmetric in x and x' for all k. However, this DeWitt-Schwinger algorithm has a singular limit as $m \to 0$ (on account of "infra-red divergences" occurring in the integration of the heat kernel needed to obtain the Green's function), and, hence, without further modification, it does not provide a suitable bi-distribution H(x,x') for defining $\langle T_{ab} \rangle$.

Nevertheless, we may construct a local bi-distribution H(x,x') via the Hadamard algorithm by simply setting $w_0 = 0$. A suitable "point-splitting" prescription for defining $\langle T_{ab} \rangle$ in state ω then may be obtained as follows. Start with the two-point distribution $\langle \phi(x)\phi(x') \rangle$ in state ω. Define

$$F(x,x') = \langle \phi(x)\phi(x') \rangle - H(x,x') \qquad (4.6.11)$$

where H(x,x') is the Hadamard bi-distribution constructed by the above algorithm with $w_0 = 0$. If we attempted to simply define $\langle T_{ab} \rangle$ via eq. (4.6.5), we would find that, in a general spacetime, $\langle T_{ab} \rangle$ would fail to be conserved on account of the failure of F(x,x') to satisfy the Klein-Gordon equation in x' at fixed x. Furthermore, if $\langle 0|\phi(x)\phi(x')|0 \rangle$ in Minkowski spacetime does not correspond to a Hadamard distribution with $w_0 = 0$ (as occurs for massive fields), then this proposed prescription would yield $\langle T_{ab} \rangle \neq 0$ for the Minkowski vacuum state, although it follows from Poincaré invariance that $\langle T_{ab} \rangle$ for the Minkowski vacuum always would be of the form $\lambda\eta_{ab}$ where λ is a constant. However, the failure of the $\langle T_{ab} \rangle$ defined via eqs. (4.6.11) and (4.6.5) to be conserved can be explicitly computed and seen to be of the form $\nabla^a \langle T_{ab} \rangle = \nabla_b Q$, where Q is a local curvature term. Hence, we may modify our prescription by simply subtracting ("by hand") the local "correction term" $Q g_{ab}$ from the formula (4.6.5) for $\langle T_{ab} \rangle$, thereby producing a prescription which yields a conserved $\langle T_{ab} \rangle$. Furthemore, the arbitrary constant in Q (undetermined from $\nabla_b Q$) can be chosen so as to make $\langle T_{ab} \rangle = 0$ for the Minkowski vacuum state in Minkowski spacetime. (An explicit calculation of Q for the case of a conformally invariant scalar field can be found in Wald (1978a); note that the presence of this "correction term" in the prescription for $<T_{ab}>$ gives rise to the existence of a "trace anomaly" for the conformally invariant field, i.e., we have $<T_a{}^a> = -4Q \neq 0$ even though, classically, $T_a{}^a$ vanishes identically for any conformally invariant field.) Thus, by modifying the naive point-splitting prescription by inclusion of the above "correction term", we obtain a prescription for defining $<T_{ab}>$ which satisfies properties (3) and (4). This prescription also manifestly satisfies property (1), and, as already indicated above, the local nature of this prescription guarantees that property (2) holds. Thus, we have established the existence of a prescription satisfying all of the desired properties. Note that since only two derivatives of F are taken prior to the coincidence limit, only the $j = 0, 1$ terms in the series for V and W survive, so in the prescription, one can truncate these series after the first two terms, thereby avoiding the necessity of confronting any issues of convergence of the series.

However, clearly this prescription does not define $<T_{ab}>$ for all states ω, but just those states for which F, eq. (4.6.11), is a smooth (or sufficiently differentiable) function. This restriction appears to be entirely reasonable from a physical point of view. Namely, as already noted above, in Minkowski spacetime the singular part of $H(x,x')$ is the same as that of $<0|\phi(x)\phi(x')|0>$. Thus, in curved spacetime, one may view the short-distance-singularity-structure of $H(x,x')$ as being "as close as possible" to that of the two-point distribution of the Minkowski vacuum. Now, the short-distance-singularity-structure of $<\phi(x)\phi(x')>$ measures the "ultraviolet behavior" of the state of the field; roughly speaking, it is governed by the state of the "high-frequency" oscillators in the infinite collection of harmonic oscillators which comprise the field. Thus, if the short-distance-singularity-structure of $<\phi(x)\phi(x')>$ is "close" to that of the Minkowski vacuum, this corresponds to having the high-frequency oscillators of the quantum field being "close" to lying in their ground state. It is only in this circumstance that one would expect to have a well defined, finite, expected stress-energy tensor of the quantum field.

On these grounds, it seems reasonable to impose as an additional requirement for the physical acceptability of a state, ω, that its two-point function, $<\phi(x)\phi(x')>$, exist and have short-distance-singularity-structure of the Hadamard form (4.6.8). (It also is would be reasonable to require $<\phi(x)\phi(x')>$ to have no "additional singularities" at "large spacelike separation" of x and x', as defined precisely in Kay and Wald (1991). However, it turns out that the positivity condition (4.5.4) actually implies the absence of "additional singularities" for a state with Hadamard short-distance-singularity-structure (Radzikowski 1992).) A state satisfying this condition is called a *Hadamard state*. Our results above may be summarized by saying that $<T_{ab}>$ is well defined—up to the above conserved local curvature term ambiguity—and nonsingular for all Hadamard states. Conversely, $<T_{ab}>$ should be singular in any non-Hadamard state. Note that if a representation of the Weyl algebra on a Hilbert space, $\mathfrak{F}$, contains a dense subspace of Hadamard states, then T_{ab} is well defined as a symmetric, quadratic form on $\mathfrak{F}$. However, even in this case, the above considerations do not guarantee that this quadratic form arises from a self-adjoint operator on $\mathfrak{F}$.

Thus, we propose that a necessary condition for a state, ω. to be physically acceptable is that it be a Hadamard state. It is essential for the viability of this proposal that every globally hyperbolic spacetime admit a wide class of Hadamard states. That this is indeed the case is a consequence of the following two results:

First, for a massive Klein-Gordon field in any static, globally hyperbolic spacetime satisfying condition (4.3.2) (so that the construction of section 4.3 is applicable) the static vacuum state is a Hadamard state (Fulling et al. 1981). It then follows immediately that in the Fock space associated with this static vacuum state, a dense set of vector states (namely, all states having only finitely many particles with associated "mode functions" which are smooth) satisfy the Hadamard condition. Thus, there exist a wide class of Hadamard states in static spacetimes. Second, it can be shown that the Hadamard condition is preserved under dynamical evolution. More precisely, if a state, ω, satisfies the Hadamard condition in a causal normal neighborhood (defined in Kay and Wald (1991)) of any Cauchy surface Σ, then it satisfies the Hadamard condition throughout the spacetime (Fulling, et al. 1978; Kay and Wald 1991).

To prove existence of a wide class of Hadamard states in an arbitrary globally hyperbolic spacetime (M, g_{ab}), we note first that we may deform (M, g_{ab}) in the manner described at the end of section 4.3 to make it static in the past, and such that eq. (4.3.2) holds. (If necessary, we also "deform" m to make it nonzero in the past.) On this deformed spacetime, we may use the static construction to define a wide class of states which satisfy the Hadamard condition in the past of (and, hence, throughout) the deformed spacetime. Then, by identifying states in the deformed spacetime with states in the original spacetime via their behavior in the future, we obtain, in turn, a wide class of Hadamard states on (M, g_{ab}). Thus, every globally hyperbolic spacetime admits a wide class of Hadamard states.

Interestingly, the myriad of unitarily inequivalent representations of the quantum field algebra obtained by the constructions of section 4.2 is greatly reduced if one restricts consideration to constructions associated with vacuum states which satisfy the Hadamard condition. In particular, for a "closed universe" (i.e., in a spacetime with a compact Cauchy surface) all Hadamard vacuum states define unitarily equivalent representations. To prove this, we appeal to the result of Fulling, et al (1981) that for a closed universe which is static in the past and in the future, the S-matrix exists, i.e., the "in" and "out" quantum field theory constructions are unitarily equivalent. The same argument establishes that for a closed universe which is static in the past, the quantum field theory construction resulting from any Hadamard vacuum state is unitarily equivalent to that arising from the "in" vacuum state. Now, suppose we are given an arbitrary (globally hyperbolic) closed universe, (M, g_{ab}), and two Hadamard vacuum states, ω_1 and ω_2. As in the previous paragraph, we may deform (M, g_{ab}) to make a globally hyperbolic spacetime (M', g'_{ab}) which is static in the past. Again, we may define corresponding Hadamard vacuum states ω_1' and ω_2' in the deformed spacetime. However, by the above result, the quantum field theory constructions associated with ω_1' and ω_2' are each unitarily equivalent to that associated with the "in" vacuum state, and, hence, are unitarily equivalent to each other. This, in turn, implies that the quantum field theory constructions associated with ω_1 and ω_2 on the original spacetime are unitarily equivalent, as we desired to show.

Thus, in the case of a closed universe, it does not appear to be essential to adopt the algebraic approach: Although there does not appear to exist any preferred "vacuum state" or preferred notion of "particle" there does exist a natural unitary equivalence class of quantum field theory constructions, namely, the ones associated with all Hadamard vacuum states. It seems plausible that all physically relevant states can be realized as density matrices on the Fock space of any Hadamard vacuum state. There does not appear to be any necessity ever to consider states lying outside of this single Hilbert space.

The fact that the quantum field theory constructions associated with Hadamard vacuum states in a closed universe are all unitarily equivalent can be understood in more intuitive terms in the following manner. The Stone-von Neumann theorem shows that the phenomenon of unitary inequivalence of representations is associated with the presence of infinitely many degrees of freedom. Now, all but finitely many degrees of freedom of a quantum field can be viewed as corresponding to "far-infrared" and "far-ultraviolet" modes. The Hadamard condition may be viewed as putting a tight restriction on the quantum state of the "far-ultraviolet" modes. On the other hand, for a closed universe, there do not exist any "far-infrared" modes. Hence, it should not be surprising that for Hadamard vacuum states in a closed universe, the situation with regard to quantum field theory constructions should be much like the case of finitely many degrees of freedom.

In the case of a spacetime with a noncompact Cauchy surface, the above considerations strongly suggest that all quantum field constructions arising from Hadamard vacuum states should be "locally indistinguishable", since the unitary inequivalence of different constructions should manifest itself only with respect to the "far-infrared" modes. This is indeed true in the following sense: Let ω_1 and ω_2 be Hadamard vacuum states, and let $\mathcal{O}$ be an open region with compact closure. Then Verch (1994) has shown that the restriction to $\mathcal{A}_{\mathcal{O}}$ of any density matrix state in the quantum field theory construction associated with ω_1 coincides with the restriction to $\mathcal{A}_{\mathcal{O}}$ of some density matrix state in the quantum field theory construction associated with ω_2. Consequently, the states in unitarily inequivalent quantum field theory constructions arising from different Hadamard vacuum states can be distinguished only by making measurements of observables over unbounded regions of spacetime. Note that—in contrast to the discussion following theorem 4.5.2—here we cannot distinguish between the states arising in the different constructions even if we make perfect measurements to perfect accuracy, provided only that we consider only observables associated with a bounded (i.e., compact closure) spacetime region. Note also that this holds not only for observables lying in the algebra $\mathcal{A}$ but also for all other observables (like the stress energy tensor) which depend locally on the state in the manner of property (2) in the above list of axioms for $<T_{ab}>$.

We turn, now, to a discussion of the "back-reaction" of the quantum field upon the spacetime geometry. Until this point, we have treated the spacetime geometry as given, and we have developed and analyzed the theory of a quantum field propagating in a fixed background spacetime. However, it is clear on physical grounds that the quantum field must have a "back-reaction" effect upon the spacetime geometry. One would expect the quantum field to be coupled to gravity via its stress-energy tensor, T_{ab}. In appropriate circumstances—in particular, if the fluctuations in T_{ab} are sufficiently small compared with T_{ab} and curvatures are small compared with the Planck scale—it seems plausible that one could continue to treat the spacetime geometry classically, and that the "back-reaction" effects would be governed by the semiclassical Einstein equation,

$$G_{ab} = 8\pi <T_{ab}> \tag{4.6.12}$$

In other words, semiclassically, the spacetimes, (M, g_{ab}), and quantum field states, ω, on (M, g_{ab}) which are dynamically possible should be the ones satisfying eq. (4.6.12). Indeed, eq. (4.6.12) can be formally derived from a full quantum theory of gravity in the "1/N" approximation, wherein one assumes the presence of N scalar fields (each coupled to gravity with coupling proportional to 1/N) and then takes the limit as $N \to \infty$. Equation (4.6.12) (with an additional "graviton contribution" to $<T_{ab}>$) also can be formally derived as the lowest order ("one-loop") perturbative correction to the expectation value of the metric in an expansion about a background classical solution. It seems highly plausible that eq. (4.6.12) will yield an adequate approximation in circumstances where the back-reaction effects are locally small—even if they lead to large, long-term, cumulative effects as occurs in black hole evaporation (see section 7.3). However, a determination and justification of the precise range of validity of eq. (4.6.12) (or even the analogous semiclassical Maxwell equation $\nabla^a F_{ab} = -4\pi <j_b>$ in quantum electrodynamics) has not yet been given.

Unfortunately, there are three serious difficulties which confront any attempt to calculate semiclassical back-reaction effects. First, as noted above, there is a two-parameter ambiguity in the definition of $<T_{ab}>$ corresponding to the addition of the two conserved local curvature terms which are quadratic in the curvature. At least one of these parameters is fundamentally undeterminable from quantum field theory in curved spacetime—i.e., it can be determined only from a more complete quantum theory of gravity or from experiment—as can be seen from the following argument (Wald 1978a): The "point-splitting" prescription given above is not scale-invariant: If one replaces g_{ab} by $\lambda^2 g_{ab}$, the stress-energy tensor does not scale homogeneously; rather, an additional term of the form $\ln\lambda$ times a conserved local curvature term quadratic in curvature also appears. (This arises on account of the presence of the non-scale-invariant "$\ln\sigma$" term in H, eq. (4.6.8). The precise linear combination of the two independent conserved, quadratic, local curvature terms which appear in this scaling law depends upon the quantum field under consideration.) Conversely, if one were presented with the correct formula for $\langle T_{ab}\rangle$, one could compare it to what one obtains from the above "point-splitting" prescription to determine a fundamental length scale in the theory. However, for the case of a massless field, no fundamental length scale is present in the formulation of quantum field theory in curved spacetime. Therefore, any argument restricted to the context of quantum field theory in curved spacetime cannot possibly uniquely determine $\langle T_{ab}\rangle$. Of course, one could simply postulate that fundamental length scale has a particular value—say, the Planck length, ℓ_p. However, any dispute concerning, say, whether the correct value is ℓ_p or $12\pi^2\ell_p$ would be resolvable only by appeal to a more complete theory or to experiment.

The second difficulty arising in the calculation of semiclassical back-reaction effects is that eq. (4.6.12) is of a "higher derivative" character than the classical Einstein equation. Namely, the classical Einstein equation is of second order in derivatives of the metric, but $\langle T_{ab}\rangle$ contains terms of fourth order in derivatives of the metric. (In addition, except in some simple, special cases, $\langle T_{ab}\rangle$ is a highly nonlocal functional of the metric.) This aspect of the semiclassical Einstein equation is analogous to the nature of the equations of motion of a point charge in classical electrodynamics when the radiation reaction term is included. As in the latter case, the semiclassical Einstein equation will admit new (presumably spurious) solutions, many of which have a "runaway" character, wherein—starting from seemingly reasonable initial conditions—the curvature blows up on scales of the Planck time. Thus, some additional criterion which distinguishes between the physically relevant and spurious solutions is required. A proposal in this regard has been given recently by Simon (1990).

The third difficulty is a practical one rather than one of principle: It is very difficult to compute $<T_{ab}>$. Even for spacetimes with a great deal of symmetry (such as Schwarzschild spacetime) and even for particularly simple states of the quantum field which share this symmetry (such as the "Hartle-Hawking vacuum state" in Schwarzschild spacetime—see section 5.3 below) it is a formidable task to obtain $<T_{ab}>$ even by numerical approximation. It would be far more difficult to calculate $<T_{ab}>$ in the much more interesting case of a dynamically evolving spacetime.

On account of these difficulties, it does not appear likely that much progress will be made in the foreseeable future toward computing detailed back-reaction effects in spacetimes of physical interest, such as spacetimes describing an evaporating black hole (see section 7.3). Interestingly, however, the above difficulties related to the properties of $<T_{ab}>$ are greatly alleviated in two spacetime dimensions, although some problems related to difficulty (2) remain. As will be discussed briefly at the end of section 7.3, it is possible that meaningful semiclassical back-reaction calculations can be performed in some two-dimensional "toy models" of gravitational theory involving a classical metric and dilaton field together with quantum scalar fields.

4.7 Other Linear Fields

In this section we briefly describe the generalization to other linear fields of our formulation of the quantum field theory of a real Klein-Gordon scalar field.

It should be clear from our discussion that the only essential structure of the classical theory used in the formulation of the quantum theory of a Klein-Gordon field was the real, symplectic vector space structure, $(\mathcal{S}, \Omega)$, of the classical solutions. Consequently, our formulation generalizes without essential change to any real, linear, bosonic field whose classical field equations (a) have a well posed initial value formulation and (b) are derivable from a Lagrangian. (The latter condition ensures the presence of a symplectic structure; see Lee and Wald (1990).) However, it should be borne in mind that the requirement that the classical field equations have a well posed initial value formulation in curved spacetime is a highly nontrivial restriction: The straightforward generalization to curved spacetime of the standard spin–s field equations in flat spacetime do not admit a well posed initial value formulation for $s > 1$ (see, e.g., Wald (1984a)).

As a concrete illustration of how quantum field theory in curved spacetime can be formulated for other real, linear bosonic fields, consider a Maxwell field in a curved, globally hyperbolic spacetime (see also Dimock, 1992). A classical Maxwell field is a connection on a U(1) bundle over spacetime, M; equivalently, a classical Maxwell field is an equivalence class of vector potentials, A_a, on spacetime, where A_a and A'_a are equivalent if they differ by a gauge transformation, i.e., $A_a - A'_a = \nabla_a \chi$ for some function χ on M. It should be noted that by working only with "gauge equivalence classes" of vector potentials, we, in effect, eliminate the constraints which occur in the usual manner in which the Hamiltonian formulation of Maxwell theory is presented.

Maxwell's equations are simply $\nabla^a F_{ab} = 0$, where $F_{ab} \equiv \nabla_a A_b - \nabla_b A_a$. These equations have a well posed initial value formulation (in the sense described in problem 2 of chapter 10 of Wald (1984a)). We define $\mathcal{S}$ to be the vector space of solutions to Maxwell's equations such that F_{ab} has compact support on a Cauchy surface Σ. (This latter condition is both gauge-independent and independent of the choice of Σ.) The symplectic structure on $\mathcal{S}$ naturally arising from the Lagrangian formulation of Maxwell theory is

$$\Omega(A_{1a}, A_{2a}) = \int_\Sigma [F_1{}^{ab} A_{2a} - F_2{}^{ab} A_{1a}]\, n_b \sqrt{h}\, d^3x \qquad (4.7.1)$$

(see, e.g., Lee and Wald (1990)). It may then be verified that—by virtue of Maxwell's equations—Ω is both gauge independent and independent of the choice of Cauchy surface Σ. Thus, Ω is well defined (and nondegenerate) on $\mathcal{S}$. The remainder of our construction of the quantum theory of a scalar field may now be taken over directly, except for the following small modification: In lemma 3.2.1 and its generalization to curved spacetime, the scalar "test function" f is replaced by a "test vector field" f^a which, in addition, must satisfy $\nabla_a f^a=0$, since solutions to Maxwell's equations with "source" f^a will exist only when $\nabla_a f^a=0$. This has the consequence that the field operator, $\hat{A}_a$, analogous to eq. (4.2.9) can be "smeared" only with divergence-free test vector fields.

We now consider the generalization of our construction of quantum field theory in curved spacetime to complex, linear bosonic fields. In many cases (such as that of a complex Klein-Gordon scalar field), the field equations in curved spacetime will be invariant under a complex conjugation. In that case, the real and imaginary parts of the field can be treated as decoupled real fields and the quantum field theory then can be formulated by straightforward application of the prescription for real fields. However, in more general cases—such as a charged Klein-Gordon scalar field with an external electromagnetic field also present—the equations are complex, and the real and imaginary parts of the field are coupled. The quantum field construction analogous to that of section 4.2 can be formulated in the general case as follows: First, the classical symplectic structure (which arises from a Lagrangian formulation of the theory) now is most naturally viewed as a complex bilinear map $\Omega:\overline{\mathcal{S}}\times\mathcal{S}\to\mathbb{C}$. Let $\mu:\overline{\mathcal{S}}\times\mathcal{S}\to\mathbb{C}$ denote a (complex) inner product on $\mathcal{S}$ satisfying the analog of eq. (4.2.7), i.e.,

$$\mu(\overline{\psi}_1,\psi_1) = \frac{1}{4}\ \underset{\psi_2\neq 0}{\text{l.u.b.}}\ \frac{[\Omega(\overline{\psi}_1,\psi_2)]^2}{\mu(\overline{\psi}_2,\psi_2)} \tag{4.7.2}$$

We complete $\mathcal{S}$ in the inner product μ to obtain a complex Hilbert space, $\mathcal{S}_\mu$, and we construct the Hilbert subspace $\mathcal{H}\subset\mathcal{S}_\mu$ exactly as in section 3.2 above. Similarly, the inner product $\overline{\mu}$ on the vector space $\overline{\mathcal{S}}$ with symplectic structure $\overline{\Omega}:\mathcal{S}\times\overline{\mathcal{S}}\to\mathbb{C}$ defines a Hilbert subspace $\mathcal{H}'\subset\overline{\mathcal{S}}_\mu$. ($\mathcal{H}'$ is the orthogonal complement to $\overline{\mathcal{H}}$ in $\overline{\mathcal{S}}_\mu$.) Elements of $\mathcal{H}$ are called "particle states", and elements of $\mathcal{H}'$ are called "antiparticle states". We take the Hilbert space of the theory to be

$$\mathcal{F}_s(\mathcal{H}\oplus\mathcal{H}') \cong \mathcal{F}_s(\mathcal{H})\otimes\mathcal{F}_s(\mathcal{H}') \tag{4.7.3}$$

As before, let $K:\mathcal{S}\to\mathcal{H}$ denote the projection map (in the inner product 2μ) from $\mathcal{S}$ to $\mathcal{H}$. Let $K':\overline{\mathcal{S}}\to\mathcal{H}'$ denote the similar projection map to $\mathcal{H}'$. The construction of the theory is completed by defining, for each $\psi\in\mathcal{S}$, the Schrodinger operator corresponding to the classical observable $\Omega(\overline{\psi},\cdot)$ by

$$\hat{\Omega}(\overline{\psi},\cdot) = ia(\overline{K\psi}) - ia^\dagger(K'\,\overline{\psi}) \tag{4.7.4}$$

(Note that $\hat{\Omega}(\overline{\psi},\cdot)$ now is not self-adjoint, corresponding to the fact that the classical observable $\Omega(\overline{\psi},\cdot)$ now is not real.) The (very minor) differences occurring in the analysis of the unitary equivalence of different constructions can be found in Wald (1979a). Again, the algebraic structure given by the commutation relations of the fundamental observables (4.7.4) is the same for all constructions, and the theory can be formulated via the algebraic approach in the same manner as for the real field.

Unlike bosonic fields, there is no "classical limit" of a fermion field, and the formulation of the quantum theory of a fermion field does not correspond simply to the quantum theory of a collection of ordinary harmonic oscillators. Nevertheless, the mathematical structure of the quantum theory of a fermion field is very closely analogous to that of the bosonic case. Consider, first, the construction of the quantum theory of a real, linear fermion field—such as a Majorana neutrino—in curved spacetime. For such a theory, there is one crucial difference occurring at the classical level from the case of a real, linear boson field: The vector space, $\mathcal{S}$, of classical solutions to the fermion field equations (say, with smooth, compact support initial data) has defined on it a natural symmetric, positive definite inner product, $\Lambda:\mathcal{S}\times\mathcal{S}\rightarrow\mathbb{R}$, rather than an antisymmetric, symplectic form, Ω. As in the bosonic case, we proceed by complexifying $\mathcal{S}$ to obtain a space denoted $\mathcal{S}^{\mathbb{C}}$. We may extend the action of the real inner product Λ from $\mathcal{S}$ to $\mathcal{S}^{\mathbb{C}}$ so that it defines a complex inner product on $\mathcal{S}^{\mathbb{C}}$. We then complete $\mathcal{S}^{\mathbb{C}}$ in this inner product to obtain a complex Hilbert space, $\mathcal{S}_{\Lambda}{}^{\mathbb{C}}$. Note that, unlike the bosonic case, there is no need to choose an analog of μ for this step. Nevertheless, a choice analogous to the choice of μ must be made to define the "one-particle Hilbert space", $\mathcal{H}$, of the quantum field theory construction. We choose $\mathcal{H}$ to be any complex subspace of $\mathcal{S}_{\Lambda}{}^{\mathbb{C}}$ which satisfies the analog of properties (ii) and (iii) of section 2.3. More precisely, the choice of $\mathcal{H}$ is restricted only by the requirement that it and its complex conjugate space, $\overline{\mathcal{H}}$, span $\mathcal{S}_{\Lambda}{}^{\mathbb{C}}$ and be orthogonal to each other in the inner product Λ. We define the inner product on $\mathcal{H}$ to be simply the restriction of Λ to $\mathcal{H}$. The Hilbert space of the quantum field theory then is taken to be the antisymmetric Fock space, $\mathcal{F}_a(\mathcal{H})$, constructed from $\mathcal{H}$ (see Appendix A.2). For each $\psi\in\mathcal{S}$, the operator $\hat{\Lambda}(\psi,\cdot)$ on $\mathcal{F}_a(\mathcal{H})$ corresponding to the classical quantity $\Lambda(\psi,\cdot)$ is then taken to be

$$\hat{\Lambda}(\psi,\cdot) = a(\overline{K\psi}) + a^{\dagger}(K\psi) \tag{4.7.5}$$

where $K: \mathcal{S}_\Lambda{}^{\mathbb{C}} \rightarrow \mathcal{H}$ and $\overline{K}: \mathcal{S}_\Lambda{}^{\mathbb{C}} \rightarrow \overline{\mathcal{H}}$ are the projection maps from $\mathcal{S}_\Lambda{}^{\mathbb{C}}$ onto $\mathcal{H}$ and $\overline{\mathcal{H}}$, respectively, and a and $a^\dagger$ denote the annihilation and creation operators on $\mathcal{F}_a(\mathcal{H})$ (defined by the analogs of eqs. (A.3.4) and (A.3.5), with antisymmetrization replacing symmetrization). The analysis of the unitary equivalence of different constructions then proceeds in close parallel with the bosonic case (see Wald (1979a) for details). It is noteworthy here that the sign changes in this analysis resulting from basing the construction upon a real inner product space, $(\mathcal{S}, \Lambda)$, rather than a real, symplectic vector space, $(\mathcal{S}, \Omega)$, are precisely compensated by the choice of antisymmetric rather than symmetric Fock space.

Equation (4.7.5) gives rise to "canonical anticommutation relations" for the fermion field operator, which have the same algebraic structure for all constructions. Consequently, the theory may be formulated via the algebraic approach using the "canonical anticommutation algebra" in parallel with the "canonical commutation algebra" used in the bosonic case. The one notable difference occurring here is that a and $a^\dagger$ are bounded operators on $\mathcal{F}_a(\mathcal{H})$, so the operators $\hat{\Lambda}(\psi, \cdot)$ also are bounded. Hence, one can work directly with the algebra of the operators (4.7.5), i.e., there is no need to exponentiate these operators in the manner done to construct the Weyl algebra in the bosonic case.

Finally, the further modifications needed to treat a complex fermion field parallel the bosonic case discussed above; see Dimock (1982) for the case of a Dirac field.

5 The Unruh Effect

As our first application of quantum field theory in curved spacetime, we return to Minkowski spacetime! In Minkowski spacetime, any one-parameter group of Lorentz boost isometries has orbits which are timelike in a globally hyperbolic region. Hence, we may view this region as a spacetime in its own right and perform the quantum field construction of section 4.3, using Lorentz boosts as the notion of "time translations". We then may compare this construction to the standard quantum field construction for all of Minkowski spacetime given in chapter 3. In particular, we may attempt to express the standard Minkowski vacuum state, $|0>_M$, as a state in the folium of the new construction. As we shall see in section 5.1, when we do so, we obtain the remarkable conclusion—known as the "Unruh effect"—that the Minkowski vacuum corresponds to a thermal state in the new construction. In physical terms, this means that when the field is in the standard vacuum state, $|0>_M$, an accelerating observer will "feel himself" to be immersed in a thermal bath of particles.

The key geometrical structure in the Minkowski spacetime analysis which gives rise to the thermal state property of $|0>_M$ is the bifurcate Killing horizon generated by the Lorentz boost isometries. The geometry of bifurcate Killing horizons is studied in section 5.2. This will give us the tools needed to generalize the Unruh effect to curved spacetimes possessing a bifurcate Killing horizon, which we shall do in section 5.3.

Although, as we shall emphasize in section 5.3, the Unruh effect is both physically and mathematically distinct from the "Hawking effect" of spontaneous particle creation by black holes, these two effects are very closely related. Indeed, the Unruh effect was discovered in an attempt to gain more insight into the nature of the Hawking effect. The mathematical machinery developed here in the course of our analysis of the Unruh effect will play an important role in our analysis of the Hawking effect, given in chapter 7.

5.1 The Unruh Effect in Flat Spacetime

Let us reconsider the construction of quantum field theory in Minkowski spacetime from the perspective of the general prescription for constructing quantum field theory in an arbitrary, globally hyperbolic, stationary spacetime (see section 4.3). Let ξ^a denote an "ordinary time translation" Killing field, corresponding to the "time direction" of some global family of inertial observers. Then eq. (4.3.2) holds, so if $m > 0$ we may apply the stationary quantization prescription of section 4.3. (If $m = 0$ this construction also works for spacetime dimension greater than two; however, infra-red divergences occur when one attempts to apply this construction to two-dimensional Minkowski spacetime.) The result of applying this prescription, of course, is precisely the standard quantum field theory that we previously constructed in section 3.2.

In fact, Minkowski spacetime possesses a three-parameter family of "ordinary time translations", corresponding to the different possible choices of "time direction" for global families of inertial observers. Hence, one could repeat the above construction making a different choice of time translation Killing field. However, it is not difficult to show that the use of a different time translation Killing field gives rise to precisely the same μ, and thus the quantum field construction actually does not depend upon the choice of ξ^a (see Chmielowski (1994) for a proof of a generalization of this result). This result may be interpreted as saying that different global families of inertial observers will agree upon the "particle content" of any state of the field.

However, Minkowski spacetime also admits other isometries. In particular, consider the one-parameter group of Lorentz boost isometries generated by the Killing field

$$b^a = a[X(\partial/\partial T)^a + T(\partial/\partial X)^a] \quad (5.1.1)$$

where a is an arbitrary constant and T and X are global inertial coordinates. Figure 5.1 shows the orbit structure of these isometries. It can be seen from that figure that although b^a fails to be globally timelike, it is timelike in the "right wedge" region labeled as "I". Note also that region I is globally hyperbolic. Hence, we can construct a quantum field theory for region I—viewed as a spacetime in its own right—by the prescription of section 4.3, using b^a as the timelike Killing vector field. (Note, however, that eq.(4.3.2) is not

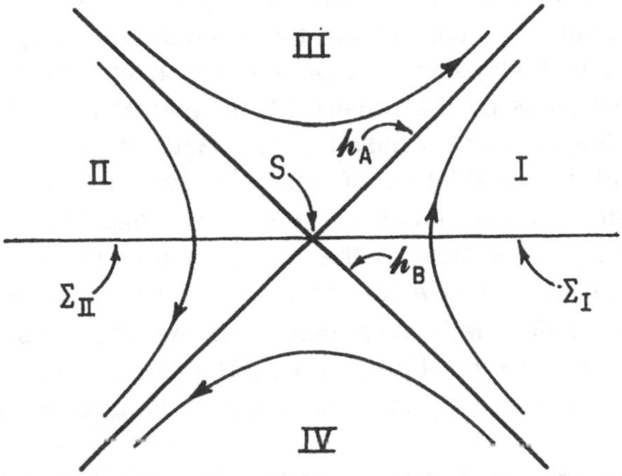

Figure 5.1. A spacetime diagram of Minkowski spacetime, showing the four "wedges" determined by the boost Killing field b^a.

satisfied, so "infra-red divergences" cannot be ruled out *a priori.*) The vacuum state, $|0\rangle_I$, for region I obtained by this construction is known as the *Rindler vacuum.* States in the associated Fock space have a natural particle interpretation for the "stationary observers" of region I, i.e., the observers following orbits of b^a. These observers all undergo uniform acceleration, although this acceleration varies from orbit to orbit. Since the acceleration of the orbit on which $b^a b_a = -1$ is a, it is conventional to view the orbits of b^a as corresponding to a family of observers naturally associated with an observer who accelerates uniformly with acceleration a. Similarly, the notion of "particles" obtained from this quantum field construction are referred to as the "particles seen by an observer who undergoes uniform acceleration a ".

The key question we wish to consider is the following: View the ordinary Minkowski vacuum state, $|0\rangle_M$, (i.e., the vacuum state of the standard quantum field construction using an "ordinary time translation" ξ^a) as an algebraic state. Its restriction to region I defines a state in that spacetime (see the discussion at the end of section 4.5). What is this state when expressed (formally) as a state in the Fock space obtained via the quantum field construction using b^a for the spacetime I? In other words, what is the "particle content" of $|0\rangle_M$ as seen by accelerated observers?

Before beginning our analysis of this question, it is worth

making two comments. First, we already know that the restriction of $|0\rangle_M$ to region I cannot equal the Rindler vacuum, $|0\rangle_I$, for the simple reason that $|0\rangle_I$ defines a pure state on the Weyl algebra, $\mathcal{A}_I$, for region I, whereas by the remark at the end of the section 4.5, the restriction of $|0\rangle_M$ defines a mixed state. Second, we have no reason to expect that the restriction of $|0\rangle_M$ actually lies in the folium of $|0\rangle_I$. However, the calculation of the formal expression for the restriction of $|0\rangle_M$ as a state in the folium of $|0\rangle_I$ is of direct interest for the reasons given below theorem 4.5.2. Thus, we shall not concern ourselves below with such issues as whether eqs. (4.4.2) and (4.4.21) are satisfied for the two constructions. (As it turns out, they are not satisfied, i.e., the restriction of $|0\rangle_M$ does not lie in the folium of $|0\rangle_I$.)

Our strategy for (formally) expressing the restriction of $|0\rangle_M$ as a state in the folium of $|0\rangle_I$ is as follows: The standard quantum field construction on all of Minkowski spacetime corresponds to choosing the one-particle-Hilbert space, $\mathcal{H}_1$, to consist of solutions which are purely positive frequency with respect to ordinary time translations generated by ξ^a. Now, let Σ be a Cauchy surface for Minkowski spacetime which passes through the "bifurcation surface", S, of fig. 5.1. Let Σ_I and Σ_{II} denote the portions of Σ lying in regions I and II, respectively (so that Σ_I and Σ_{II} are Cauchy surfaces for these regions). Consider a second quantum field construction for all of Minkowski spacetime obtained by choosing the one-particle-Hilbert space, $\mathcal{H}_2$, to be

$$\mathcal{H}_2 = \mathcal{H}_I \oplus \mathcal{H}_{II} \tag{5.1.2}$$

where $\mathcal{H}_I$ consists of solutions with initial data with support on Σ_I which are positive frequency with respect to b^a, and $\mathcal{H}_{II}$ consists of solutions with initial data with support on Σ_{II} which are positive frequency with respect to $-b^a$, i.e., $\mathcal{H}_I$ and $\mathcal{H}_{II}$ are the one-particle Hilbert spaces arising from the construction of section 4.3 applied to regions I and II, respectively. (We must use $-b^a$ rather than b^a to define "positive frequency" in region II because $-b^a$ is future-directed there.) We shall use eq. (4.4.23) to compute the action of the "S-matrix" $U:\mathcal{F}_s(\mathcal{H}_1)\to\mathcal{F}_s(\mathcal{H}_2)$ on $|0\rangle_M = |0\rangle_1 \in \mathcal{F}_s(\mathcal{H}_1)$. In this way, we shall (formally) express the Minkowski vacuum as a state in

$$\mathcal{F}_s(\mathcal{H}_2) \cong \mathcal{F}_s(\mathcal{H}_I)\otimes\mathcal{F}_s(\mathcal{H}_{II}) \tag{5.1.3}$$

By "tracing out" over the states in $\mathcal{F}_s(\mathcal{H}_{II})$ we shall obtain the desired density matrix on $\mathcal{F}_s(\mathcal{H}_I)$ corresponding to the restriction of $|0>_M$ to region I. Indeed, the intermediate stages of this calculation will yield additional results of interest: The expression yielding $U|0>_M$ as a state in $\mathcal{F}_s(\mathcal{H}_I)\otimes\mathcal{F}_s(\mathcal{H}_{II})$ will display the correlations between the particle content seen by accelerating observers in regions I and II.

The following three observations will aid us considerably in our calculation of the S-matrix. First, in terms of the global inertial coordinates T, X, Y, Z, we define advanced and retarded null coordinates U, V by

$$U = T - X, \quad V = T + X \tag{5.1.4}$$

Then it is not difficult to show that a solution lies in $\mathcal{H}_1$ (i.e., it is positive frequency with respect to T in the coordinates T, X, Y, Z) if and only if it is positive frequency with respect to U in the coordinates U, V, Y, Z or, equivalently, if and only if it is positive frequency with respect to V in these coordinates.

Second, consider the intersecting null planes, h_A and h_B, of fig. 5.l, to which b^a is normal. (As will be discussed further in the next section, these planes comprise a bifurcate Killing horizon.) Then it follows from results from the standard null initial value formulation together with the Holmgren uniqueness theorem that any solution to the Klein-Gordon equation (3.1.2) on Minkowski spacetime is uniquely determined by its restriction to $h_A \cup h_B$ (see proposition 2.5 of Kay and Wald (1991)). Indeed, it is worth pointing out that for the massless Klein-Gordon field in Minkowski spacetime of any dimension greater than two, a considerably stronger result holds: Any solution $\psi \in \mathcal{S}$ is uniquely determined by its restriction to h_A alone (or, alternatively, to h_B alone). This can be proven most simply by conformally completing Minkowski spacetime and noting that any $\psi \in \mathcal{S}$ has a smooth extension to this conformal completion (see, e.g., proposition 11.1.1 of Wald (1984a)). In the conformally completed spacetime, the null plane h_A is the past boundary—minus the single generator lying on $\mathscr{I}^+$—of a point $p \in \mathscr{I}^+$ and it also is the future boundary—minus the generator on $\mathscr{I}^-$—of a point $q \in \mathscr{I}^-$. The value of ψ on the two "missing generators" of these boundaries is determined by continuity (in dimension greater than 2). Standard results on the "null cone" initial value formulation then imply that ψ is uniquely determined both in the past of p and in the future of q. Since the union of these sets includes all of Minkowski spacetime, we obtain the claimed stronger result. (This stronger result presumably also holds in the massive case, although I am not aware of a proof.) Note that in two-dimensional Minkowski spacetime—where the above argument breaks down—the null initial value formulation for intersecting null lines (combined with the fact that there is no mathematical distinction between "spacelike evolution" and "timelike evolution") implies that initial data can be freely specified on $\mathfrak{h}_A$ and $\mathfrak{h}_B$ independently. However, a solution still is uniquely determined by its restriction to $\mathfrak{h}_A \cup \mathfrak{h}_B$, in accord with the more general uniqueness result stated at the beginning of this paragraph.

The third observation is that on $\mathfrak{h}_A$, the relation between "inertial time" V and "Killing parameter time" v (determined by $b^a \nabla_a v = 1$) is given by

$$v = \frac{1}{a} \ln |V| \qquad (5.1.5)$$

To see this, we note that in the coordinates U, V, Y, Z we have

$$b^a = a[-U(\partial/\partial U)^a + V(\partial/\partial V)^a] \qquad (5.1.6)$$

so that on the surface $\mathfrak{h}_A$ defined by U = 0, we have

$$aV\frac{\partial v}{\partial V} = 1 \qquad (5.1.7)$$

from which eq. (5.1.5) follows immediately. Similarly, on $\mathfrak{h}_B$ we have

$$u = -\frac{1}{a} \ln|U| \qquad (5.1.8)$$

where u denotes Killing parameter time on $\mathfrak{h}_B$.

The main upshot of the above observations is that in order to decompose a solution $\psi \in \mathcal{S}$ into its positive and negative frequency parts with respect to "ordinary inertial time", it suffices to evaluate ψ on $\mathfrak{h}_A$ and $\mathfrak{h}_B$ and take its Fourier transform there (with respect to V on h_A and with respect to U on $\mathfrak{h}_B$). Similarly, in order to decompose ψ into its positive and negative frequency parts with respect to "accelerating time", it suffices to evaluate ψ on $\mathfrak{h}_A$ and $\mathfrak{h}_B$ and take its Fourier transform there (with respect to v on $\mathfrak{h}_A$ and with respect to u on $\mathfrak{h}_B$). Hence, we can obtain the operators C, D needed to obtain the S-matrix simply by comparing the notions of "positive frequency part" of functions on $\mathfrak{h}_A$ and $\mathfrak{h}_B$ determined by the two types of time functions namely, "inertial time" (V or U) and "accelerating time" (v or u). In this manner, we can entirely avoid the necessity of solving the Klein-Gordon equation (3.1.2) to obtain the needed relationships.

Consider, now, a solution, $\psi_{I\omega}$, to the Klein-Gordon equation (3.1.2) which vanishes in region II and oscillates harmonically with frequency $\omega > 0$ with respect to time defined by b^a in region I. In analogy with the "plane waves" commonly used in ordinary field theory applications, such a solution will not have finite Klein-Gordon norm, i.e., it will not define a normalizable vector in $\mathcal{H}_I$. Consequently, some of the integrals written down below will not converge absolutely. This difficulty can be cured by working instead with a basis of normalized "wavepackets" of positive frequency solutions, whose frequencies are peaked sharply about ω (see Hawking (1975) or Wald (1975) for an explicit construction of such an orthonormal basis). We will pass to such a basis in our final expressions, but in order to avoid the cumbersome aspects of using wavepackets, we will simply work directly with $\psi_{I\omega}$ in our initial calculations.

Let $f_{I\omega}$ denote the restriction of $\psi_{I\omega}$ to $\mathfrak{h}_A$. Then $f_{I\omega}$ takes the form

$$f_{I\omega}(V,Y,Z) = \begin{cases} h(Y,Z)\exp(-i\omega v) & V>0 \\ 0 & V<0 \end{cases} \tag{5.1.9}$$

We wish to decompose $f_{I\omega}$ into its positive and negative frequency parts with respect to V. (This will directly yield the operators A and B of section 4.4; however, as we shall see, the relations we obtain will allow us to directly obtain C^{-1} and $\mathcal{E} = \overline{D}\,\overline{C}^{-1}$, as needed to evaluate the right side of eq. (4.4.23).) Since we have already used the letter "ω" to denote the frequency with respect to v, we use "σ" to denote frequency with respect to V. The Fourier transform of $f_{I\omega}$ with respect to V is

$$\hat{f}_{I\omega}(\sigma,Y,Z) = \frac{1}{\sqrt{2\pi}} \int_{-\infty}^{\infty} \exp(i\sigma V)\, f_{I\omega}(V,Y,Z)\, dV$$

$$= \frac{1}{\sqrt{2\pi}} h(Y,Z)\, I(\sigma) \qquad (5.1.10)$$

where

$$I(\sigma) = \int_0^\infty \exp(i\sigma V)\, \exp(-\frac{i\omega}{a} \ln V)\, dV \qquad (5.1.11)$$

where eqs. (5.1.9) and (5.1.5) were used. Note that this integral does not converge absolutely, but the corresponding integral with $f_{I\omega}$ replaced by a suitable wavepacket would converge.

Now, let $\sigma > 0$. We wish to compare $I(\sigma)$ with $I(-\sigma)$. In order to do so, we extend the "ln" function to the complex plane so that its branch cut lies on the negative real axis—and, hence, out of the way of the contour rotations we shall perform below. To evaluate $I(\sigma)$ for $\sigma >$ o, we substitute V = iy and rotate the contour of integration so as to make y real and positive. Using the relation

$$\ln V = \ln(iy) = \frac{i\pi}{2} + \ln y \qquad (5.1.12)$$

we obtain

$$I(\sigma) = i \exp(\pi\omega/2a) \int_0^\infty \exp(-\sigma y)\, \exp(-\frac{i\omega}{a} \ln y)\, dy \qquad (5.1.13)$$

To obtain a similar expression for $I(-\sigma)$, we substitute V = –iy and again rotate the contour to make y real and positive. We obtain

$$I(-\sigma) = -i \exp(-\pi\omega/2a) \int_0^\infty \exp(-\sigma y)\, \exp(-\frac{i\omega}{a} \ln y)\, dy \qquad (5.1.14)$$

Consequently, we have for all $\sigma >$ o,

$$\hat{f}_{I\omega}(-\sigma,Y,Z) = -\exp(-\pi\omega/a)\, \hat{f}_{I\omega}(\sigma,Y,Z) \qquad (5.1.15)$$

Consider, now, the "wedge reflection" isometry of Minkowski spacetime given by (T,X,Y,Z)→(–T,–X,Y,Z). Under this isometry, regions I and II are interchanged. The induced action of this isometry on an element of $\mathcal{H}_I$—i.e., a solution which is purely positive frequency with respect to b^a in region I and which vanishes in region II—

maps it into a solution in $\overline{\mathcal{H}}_{\text{II}}$. We denote by $\bar{f}_{\text{II}\omega}$ the restriction to h_A of the image of $\psi_{\text{I}\omega}$ under the induced action of this isometry. Clearly, we have

$$\bar{f}_{\text{II}\omega}(V,Y,Z) = \begin{cases} 0 & V>0 \\ h(Y,Z)\exp(-[i\omega/a]\ln|V|) & V<0 \end{cases} \tag{5.1.16}$$

The fact that $\bar{f}_{\text{II}\omega}$ is the "time reflection" of $f_{\text{I}\omega}$ immediately implies that for all σ (positive or negative) we have

$$\hat{\bar{f}}_{\text{II}\omega}(\sigma,Y,Z) = \hat{f}_{\text{I}\omega}(-\sigma,Y,Z) \tag{5.1.17}$$

Define the function F_ω on h_A by

$$F_\omega = f_{\text{I}\omega} + \exp(-\pi\,\omega/a)\,\bar{f}_{\text{II}\omega} \tag{5.1.18}$$

Then, it follows from eqs. (5.1.15) and (5.1.17) that F_ω has no negative frequency part with respect to V, i.e., for all $\sigma > o$ we have

$$\hat{F}_\omega(-\sigma, Y, Z) = 0 \tag{5.1.19}$$

Similar results hold, of course, on h_B.

As noted above, solutions of the form $\psi_{\text{I}\omega}$ do not define normalizable elements of $\mathcal{H}_{\text{I}}$. However, as also noted above, we can choose an orthonormal basis $\{\psi_{\text{I}i}\}$ of $\mathcal{H}_{\text{I}}$ such that each $\psi_{\text{I}i}$ is sharply peaked about some frequency ω_i. Let $\{\psi_{\text{II}i}\}$ denote the corresponding orthonormal basis of $\mathcal{H}_{\text{II}}$, obtained by applying the "wedge reflection" isometry to $\{\psi_{\text{I}i}\}$, followed by complex conjugation. Then, to the extent that the peaking of $\psi_{\text{I}i}$ about the frequency ω_i is very sharp, it follows from eq. (5.1.19) that the solution

$$\Psi_i = \psi_{\text{I}i} + \exp(-\pi\omega_i/a)\ \bar{\psi}_{\text{II}i} \tag{5.1.20}$$

is purely positive frequency with respect to inertial time, and thus defines an element of $\mathcal{H}_1$. By exactly similar arguments, it follows that

$$\Psi'_i = \psi_{\text{II}i} + \exp(-\pi\omega_i/a)\ \bar{\psi}_{\text{I}i} \tag{5.1.21}$$

also is purely positive frequency with respect to inertial time.

These facts are precisely what we need to calculate the S-matrix, U. By inspection, we have

$$C\Psi_i = \psi_{iI}\,,\quad C\Psi'_i = \psi_{iII} \tag{5.1.22}$$

$$D\Psi_i = \exp(-\pi\omega_i/a)\,\overline{\psi}_{iII}\,,\qquad D\Psi'_i = \exp(-\pi\omega_i/a)\,\overline{\psi}_{iI} \tag{5.1.23}$$

where $C:\mathcal{H}_1\to\mathcal{H}_2$ and $D:\mathcal{H}_1\to\overline{\mathcal{H}}_2$ were defined in section 4.4. Hence, from eqs. (5.1.22) and (5.1.23), we obtain

$$DC^{-1}\psi_{iI} = \exp(-\pi\omega_i/a)\,\overline{\psi}_{iII} \tag{5.1.24}$$
$$DC^{-1}\psi_{iII} = \exp(-\pi\omega_i/a)\,\overline{\psi}_{iI} \tag{5.1.25}$$

Since $\{\psi_{iI}\}$ and $\{\psi_{iII}\}$ jointly span $\mathcal{H}_2 = \mathcal{H}_I \oplus \mathcal{H}_{II}$, we thus have determined the operator $\mathcal{E} = \overline{D}\,\overline{C}^{-1}$. The corresponding two-particle state is

$$\varepsilon^{ab} = \sum_i \exp(-\pi\omega_i/a)\; 2(\psi_{iI})^{(a}(\psi_{iII})^{b)} \tag{5.1.26}$$

This yields—in Dirac rather than index notation—

$$U|0\rangle_1 = \prod_i \Big\{ \sum_{n=0}^{\infty} \exp(-n\pi\omega_i/a)\; |n_{iI}\rangle \otimes |n_{iII}\rangle \Big\} \tag{5.1.27}$$

where $|n_{iI}\rangle \in \mathcal{F}_s(\mathcal{H}_I)$ denotes the state corresponding to having n particles in the mode ψ_{iI} and $|n_{iII}\rangle$ denotes the state in $\mathcal{F}_s(\mathcal{H}_{II})$ with n particles in the mode ψ_{iII}. Equation (5.1.27) is the desired representation of the ordinary Minkowski vacuum state $|0\rangle_1$ as a state in $\mathcal{F}_s(\mathcal{H}_2)$. The expression is formal in that the right side of eq. (5.1.27) does not define a normalizable element of $\mathcal{F}_s(\mathcal{H}_2)$, i.e., the two quantum field constructions are unitarily inequivalent (as was first shown by Fulling (1973)). Thus, as already mentioned above, eq. (5.1.27) should be interpreted in the manner discussed below theorem 4.5.2.

A striking feature of eq. (5.1.27) is the correlations occurring in the "particle content" of the Minkowski vacuum as seen by accelerating observers in regions I and II. We see that n particles in mode

ψ_{i1} will be present in region I if and only if exactly n particles in mode ψ_{iII} are present in region II. Of course, such correlations are not, in any sense, "new physics." In the standard formulation of the quantum field theory, they correspond to the fact that the correlation functions of the quantum field in the Minkowski vacuum state are nonvanishing at spacelike separated events. All of the correlations in physical observations made by accelerating observers in regions I and II can be derived without representing the Minkowski vacuum in the form (5.1.27). However, eq. (5.1.27) displays these correlations in a very graphic way.

To obtain the density matrix which formally represents the restriction of the Minkowski vacuum to region I, we take the tensor product of the right side of eq. (5.1.27) with its dual, and then take the partial trace of the result with respect to a basis of $\mathcal{F}_s(\mathcal{H}_2)$. We immediately obtain

$$\rho = \prod_i \{ \sum_{n=0}^{\infty} \exp(-2\pi n\omega_i/a) \, |n_{iI}\rangle\langle n_{iI}| \} \qquad (5.1.28)$$

where, again, Dirac notation has been used. Now, $n\omega_i$ is precisely the energy of the state $|n_{iI}\rangle$ "as seen by the accelerating observers". More precisely, the notion of "time translation symmetry" defined by b^a in region I defines a conjugate notion of energy in that region, and the value—above the ground state $|0\rangle_I$—of this energy in the state $|n_{iI}\rangle$ is $n\omega_i$. Hence, the right side of eq. (5.1.28) is precisely of the form of the thermal density matrix $\exp(-H_I/T)$, where H_I is the Hamiltonian for region I (conjugate to the notion of time translation defined by b^a) and where

$$T = a/2\pi \qquad (5.1.29)$$

Thus, the restriction of $|0\rangle_M$ to region I is precisely a thermal state with temperature given by eq. (5.1.29). This result—known as the "Unruh effect" (Unruh 1976)—can be interpreted as saying that an accelerating observer "feels himself immersed in a thermal bath of particles" at temperature (5.1.29). Note that in cgs units, eq. (5.1.29) becomes

$$(T/1\,^\circ\mathrm{K}) \approx (a/10^{21}\ \mathrm{cm/sec^2}) \qquad (5.1.30)$$

so the effect is extremely small for accelerations achievable by macroscopic bodies.

The Unruh effect may appear paradoxical to readers who are used to thinking that quantum field theory is, fundamentally, a theory of "particles", and that the notion of "particles" has objective significance. How can an accelerating observer assert that "particles" are present in region I when any inertial observer would assert that, "in reality", all of Minkowski spacetime is devoid of particles? Which of these two observers is "correct" in his assertion? The answer, of course, is that both observers are correct: It simply happens that the natural notion of "particles" defined by accelerating observers (convenient for characterizing the behavior of "particle detectors"—like the model of section 3.3—which are "time translationally invariant" with respect to b^a) differs from the natural notion of particles defined by inertial observers (convenient for characterizing the behavior of detectors which are invariant under ordinary time translations). No paradox arises when one views quantum field theory as, fundamentally, being a theory of local field observables, with the notion of "particles" merely being introduced as a convenient way of labeling states in certain situations.

Nevertheless, the same physical predictions must be obtained whether one labels the states of the quantum field theory via the natural labeling of the accelerating observer or that of the inertial observer. Now, when the field is in the Minkowski vacuum state, $|0>_1$,—and, thus, is in the thermal density matrix (5.1.28) in the natural labeling given by the accelerating observer—there is a nonzero probability that a particle detector carried by the accelerating observer will make a transition to an excited state. The accelerating observer, of course, would describe this process as being simply the result of the absorption of a "particle" by his detector. The inertial observer must also see this transition in the state of the accelerating detector (as well as the accompanying change in the state of the field), and it is instructive to analyze how he would explain what has occurred. This can be worked out explicitly for the simple model particle detector considered in section 3.3. The result is that the inertial observer would describe this process as the emission of a "particle" by the "detector", accompanied by a change in the state of the detector due to "radiation reaction". Further details of the inertial description of this process can be found in Unruh and Wald (1984).

It should be emphasized that the detailed properties of the "thermal bath seen by accelerating observers" may, in principle, differ significantly from the corresponding properties of an ordinary (i.e., inertial) thermal bath at the same temperature. This point is well illustrated by the following example: Consider two ordinary boxes of gas, the first one undergoing inertial motion (i.e., freely falling in outer space) and the second one placed on a table in the gravitational field of the Earth. There is a well defined notion of a "thermal equilibrium state at temperature T" for each of these boxes. However, for the first box, the density of the gas is uniform throughout the box when it is in its thermal equilibrium state, whereas for the second box, the density of gas in thermal equilibrium is slightly higher at the bottom of the box. Many other properties of the thermal equilibrium states of the two boxes of gas are measurably different. In the same way, there is no reason why, for example, the two-point correlation function for the Minkowski vacuum evaluated along the world line of an accelerating observer need be equal to the two-point correlation function for an ordinary (inertial) thermal equilibrium state at temperature (5.1.29) evaluated along the world line of an inertial observer. In fact, for a scalar field, these two-point correlation functions happen to be equal, and this fact is sometimes (erroneously) regarded as a derivation of the Unruh effect. However, for higher spin fields, this equality of the correlation functions does not hold. We refer the reader to the comprehensive review of Takagi (1986) for further discussion of this issue and other aspects of the Unruh effect in Minkowski spacetime.

Finally, we comment that our demonstration that the restriction of the Minkowski vacuum state, $|0>_M$, to region I is a thermal state (with respect to the notion of time translation defined by b^a) is rather unsatisfactory from the point of view of mathematical precision. What we showed is that the formal expression for the Minkowski vacuum as a density matrix on $\mathcal{F}_s(\mathcal{H}_I)$ equals the formal expression for a thermal density matrix on $\mathcal{F}_s(\mathcal{H}_I)$. In other words, we are claiming to show that two states are equal by attempting to express them in the folium of the "stationary vacuum state" for region I, but neither of these states actually belongs to that folium! To proceed more rigorously, one must properly define the notion of a "thermal state at temperature T" for a quantum system with a Hamiltonian having continuous spectrum; one cannot define it simply as being the density matrix exp(−H/T) (viewed as an operator acting on the folium of the stationary vacuum state of section 4.3), since we do not thereby obtain a normalizable density matrix. What is needed is a characterization of what it means for an algebraic state, ω, to be a thermal equilibrium state without presuming that ω exists as a density matrix in any particular representation of the field algebra $\mathcal{A}$. Such a characterization is provided by the KMS condition (see, e.g., Bratteli and Robinson (1981)). Thus, the statement that the restriction of the Minkowski vacuum to region I is a thermal state at temperature (5.1.29) can be formulated rigorously by saying that the algebraic state on $\mathcal{A}_I$ obtained by restriction of the Minkowski vacuum satisfies the KMS condition at temperature (5.1.29) with respect to the notion of "time translation" defined by b^a.

Remarkably, as pointed out by Sewell (1982), a proof of this mathematically rigorous formulation of the Unruh effect follows directly from results obtained by Bisognano and Wichmann (1976) in work done entirely independently of (and, essentially, simultaneously with) the analysis of Unruh (1976); see Kay (1985) for further discussion. Indeed, the Bisognano-Wichmann theorem holds in the much more general context of axiomatic quantum field theory, thus showing that the results we have obtained above are not restricted merely to the case of linear quantum fields; see also Unruh and Weiss (1984) for an argument for this latter conclusion based upon a path integral approach.

5.2 Killing Horizons

The geometrical structure which played the key role in the derivation of the Unruh effect in Minkowski spacetime was the pair of intersecting null planes, h_A and h_B, to which the Killing field, b^a, is normal. In this section, we shall generalize this structure to curved spacetime by defining the notion of a bifurcate Killing horizon and deriving some of its key properties. This will enable us to generalize the Unruh effect to curved spacetime in the next section.

Recall that an *isometry*, $i: M \to M$, on a spacetime, (M, g_{ab}), is a diffeomorphism which leaves the metric invariant. A *Killing vector field*, χ^a, is the infinitesimal generator of a one-parameter group of isometries, i_t. It satisfies

$$0 = \mathcal{L}_\chi g_{ab} = 2\nabla_{(a}\chi_{b)} \tag{5.2.1}$$

where ∇_a is the derivative operator associated with g_{ab} and $\mathcal{L}$ denotes the Lie derivative.

An important property of an isometry is that its action on a connected spacetime, (M, g_{ab}), is uniquely determined by its action on a single point $p \in M$ together with its induced action, $i^*:V_p \to V_{i(p)}$, on the tangent space, V_p, at p. To prove this, we note that isometries map geodesics to geodesics, and that a geodesic through p is uniquely determined by its tangent at p. Thus, the above information is sufficient to determine the action of i on all geodesics through p. More generally, we uniquely determine i on all "broken geodesics" through p, i.e., all curves through p which are piecewise geodesic. However, since any point $q \in M$ can be connected to p by a broken geodesic, i is uniquely determined on M.

Applying this result to a one-parameter group of isometries, i_t, we see that a Killing vector field, χ^a, is uniquely determined by its value and the value of its derivative $F_{ab} \equiv \nabla_a \chi_b = \nabla_{[a}\chi_{b]}$ at any point $p \in M$. This result may also be proven directly from the equation,

$$\nabla_c \nabla_a \chi_b = -R_{abcd}\,\chi^d \tag{5.2.2}$$

which is satisfied by any Killing field (see, e.g., Appendix C of Wald (1984a)).

Consider, now, the case of a two-dimensional manifold, M, and suppose a Killing field, χ^a, vanishes at a point $p \in M$, i.e., $\chi^a(p) = 0$. Then χ^a is determined by the value of the antisymmetric tensor $F_{ab} = \nabla_a \chi_b$ at p, which is unique up to scaling since is V_p two-dimensional. The induced action, $i_t^*:V_p \to V_{i(p)} = V_p$ of the group of isometries, i_t, on the tangent space, V_p, at p is determined by its "infinitesimal action". However, the infinitesimal action of i_t^* defines the Lie derivative with respect to χ^a, i.e., for all $v^a \in V_p$, the infinitesimal action of i_t^* takes v^a into

$$\mathcal{L}_\chi v^a = F^a{}_b\, v^b \tag{5.2.3}$$

The nature of the map on V_p defined by eq. (5.2.3) depends upon the signature of the metric. (The metric enters eq. (5.2.3) via the raising of an index of F_{ab}.) In the case of a metric of Riemannian signature, eq. (5.2.3) is identical to the action of an infinitesimal rotation in a flat, Euclidean space. It follows immediately that $i_t^*:V_p \to V_p$ is simply an ordinary rotation on a tangent space. In particular, when t takes the value t_0 corresponding to a "2π-rotation", $i_t{}^*$ is the identity transformation on V_p. By the above uniqueness result, this implies that i_t is the identity transformation on M. Consequently, all the orbits of i_t on M are closed (with period t_0), and, in a neighborhood of p, the orbits have the usual structure of rotations about the "axis" p.

On the other hand, for a metric of Lorentz signature, eq. (5.2.3) corresponds to the action of an infinitesimal Lorentz boost in a two-dimensional Minkowski spacetime. Thus, $i_t{}^*: V_p \to V_p$ is simply an ordinary Lorentz boost. It follows that the orbit structure of i_t in a neighborhood of p has the same structure as shown in fig. 5.1 above.

In $n > 2$ dimensions with a metric of either Riemannian or Lorentzian signature, exactly the same analysis applies for any Killing field χ^a which vanishes on an (n–2)-dimensional spacelike surface S. In the Lorentzian case, the pair of null surfaces $\mathfrak{h}_A$ and $\mathfrak{h}_B$ (intersecting at S) which are generated by the null geodesics orthogonal to S is called a *bifurcate Killing horizon*. A bifurcate Killing horizon locally divides the spacetime into four "wedges" I, II, III, IV in the manner shown in fig. 5.1. Note that since $i_t{}^*$ takes a null vector orthogonal to S at any $p \in S$ into a multiple of itself, it follows that i_t takes null geodesics orthogonal to S into themselves (with a possibly different affine parametrization). Consequently, χ^a must be everywhere tangent to the null geodesics orthogonal to S, i.e., χ^a is normal to both $\mathfrak{h}_A$ and $\mathfrak{h}_B$.

Since bifurcate Killing horizons occur in Lorentzian spacetimes whenever a Killing field vanishes on an (n–2)-dimensional spacelike surface, they can be expected to arise commonly in spacetimes with a high degree of symmetry. An important additional reason for their physical relevance arises from the fact that, under suitable assumptions concerning matter fields and under the assumption of analyticity, it can be shown (Hawking and Ellis 1973) that the event horizon of a stationary black hole must be a *Killing horizon*, i.e., a null surface to which a Killing field is normal. As will be discussed further in section 6.2, this Killing horizon should comprise a portion of a bifurcate Killing horizon, except in the "degenerate case" of vanishing surface gravity (see below). Thus, bifurcate Killing horizons should arise in essentially all spacetimes representing a stationary black hole.

We now introduce the notion of the surface gravity of an arbitrary (i.e., not necessarily bifurcate) Killing horizon. Let $\mathfrak{h}$ be a

Killing horizon associated with the Killing field χ^a, i.e., $\mathfrak{h}$ is a null hypersurface to which χ^a is normal. Since, by definition, we have $\chi^a\chi_a = 0$ on $\mathfrak{h}$, the vector $\nabla^b(\chi^a\chi_a)$ must be normal to $\mathfrak{h}$. Since χ^a also is normal to $\mathfrak{h}$, these vectors must be proportional. Hence, there exists a function κ on $\mathfrak{h}$—known as the *surface gravity* of $\mathfrak{h}$—which satisfies

$$\nabla^b (\chi^a\chi_a) = -2\kappa\chi^b \tag{5.2.4}$$

everywhere on $\mathfrak{h}$. By taking the Lie derivative of this equation with respect to χ^a, it follows immediately that κ must be constant along each orbit of χ^a on $\mathfrak{h}$.

A relatively straightforward calculation (see, e.g., Wald (1984a)) using Killing's equation together with the fact that $\chi_{[a}\nabla_b\chi_{c]}=0$ on $\mathfrak{h}$ (since χ^a is hypersurface orthogonal on $\mathfrak{h}$) yields the formula

$$\kappa^2 = -\frac{1}{2}(\nabla_a\chi_b)(\nabla^a\chi^b) \tag{5.2.5}$$

where evaluation on $\mathfrak{h}$ is understood. It follows from this equation that κ can be expressed as

$$\kappa = \lim\,[a(-\chi^a\chi_a)^{1/2}] \tag{5.2.6}$$

(see, e.g., Wald (1984a) for a derivation), where a denotes the magnitude of the (proper) acceleration of the orbits of χ^a in the region "outside" $\mathfrak{h}$ where χ^a is timelike, and the limit approaching the horizon is taken. (Note that $\chi^a\chi_a \to 0$ and $a \to \infty$ in this limit.) In the case of an asymptotically flat spacetime where χ^a approaches a time translation at infinity, it is natural to normalize χ^a so that $\chi^a\chi_a \to -1$ at infinity. Then $(-\chi^a\chi_a)^{1/2}$ is the "gravitational redshift factor", so eq. (5.2.6) gives κ the interpretation of being the "redshifted proper acceleration" of the orbits of χ^a near the horizon. The terminology "surface gravity" derives from this interpretation. If χ^a does not approach a time translation at infinity, a similar interpretation of κ as the "redshifted proper acceleration of zero angular momentum observers" can be given; see, e.g., Jacobson and Kang (1993) for further discussion of the various equivalent notions of surface gravity.

Above, we mentioned that κ is constant along each generator of an arbitrary Killing horizon $\mathfrak{h}$. A remarkable property of bifurcate Killing horizons is that κ also does not vary from generator to generator, so that κ is globally constant on h. To see this, we differentiate eq. (5.2.5) on the bifurcation surface S in a direction, s^a, tangent to S. We obtain

$$\kappa s^a \nabla_a \kappa = -\frac{1}{2} s^c (\nabla_c \nabla_a \chi_b)(\nabla^a \chi^b)$$

$$= \frac{1}{2} s^c R_{abc}{}^d \chi_d \nabla^a \chi^b$$

$$= 0 \qquad (5.2.7)$$

where we used eq. (5.2.2) in the second line, and we used the fact that $\chi^a = 0$ on S in the last line. Thus, κ is constant on S. Since κ also is constant along each generator of any Killing horizon, it follows that for any bifurcate Killing horizon, κ is constant on $h_A \cup h_B$. Note that no use of Einstein's equation was made in this argument, i.e., the constancy of κ holds for an arbitrary bifurcate Killing horizon in a spacetime of any dimension. As we shall discuss further in section 6.2, for four-dimensional spacetimes, Einstein's equation together with the dominant energy condition holding for matter implies the constancy of κ on any (not necessarily bifurcate) Killing horizon.

We conclude this section by showing that the relationship between "Killing parameter time" and "affine parameter time" on any bifurcate Killing horizon corresponds exactly to the relationships (5.1.5) and (5.1.8) obtained in the particular case of Minkowski spacetime. Using Killing's equation $\nabla_{(a}\chi_{b)} = 0$, we may rewrite eq. (5.2.4) as

$$\chi^a \nabla_a \chi^b = \kappa \chi^b \qquad (5.2.8)$$

which is just the geodesic equation in nonaffinely parametrized form. It follows immediately that if v is a Killing parameter on h_A — i.e., if the function v on the portion of h_A to the future of S satisfies $\chi^a \nabla_a v = 1$, then

$$V = \exp(\kappa v) \qquad (5.2.9)$$

is an affine parameter along the null geodesics tangent to χ^a which generate h_A. Continuing this affine parametrization to the past of S, we find that the relationship between "Killing time" and "affine time" on h_A is given by

$$v = \frac{1}{\kappa} \ln|V| \qquad (5.2.10)$$

which is of the same form as eq. (5.1.5), as we desired to show. Similarly, on h_B, the relationship between "Killing time" u and "affine time" U is given by

$$u = -\frac{1}{\kappa} \ln|U| \qquad (5.2.11)$$

5.3 The Unruh Effect in Curved Spacetime

Let (M, g_{ab}) be a spacetime possessing a bifurcate Killing horizon. We assume, in addition that (M, g_{ab}) is globally hyperbolic and possesses a Cauchy surface passing through the bifurcation surface, S. In that case, a global definition of the 4 "wedges" determined by the bifurcate Killing horizon (corresponding to the wedges of Minkowski spacetime shown in fig. 5.1 above) may be given as follows:

$$\text{Region I} = I^-(h_A) \cap I^+(h_B) \qquad (5.3.1a)$$
$$\text{Region II} = I^+(h_A) \cap I^-(h_B) \qquad (5.3.1b)$$
$$\text{Region III} = J^+(S) \qquad (5.3.1c)$$
$$\text{Region IV} = J^-(S) \qquad (5.3.1d)$$

It follows that these regions are disjoint, except for the intersection of regions III and IV on S (see Kay and Wald (1991)). Furthermore, by the arguments given in the previous section (see the discussion below eq. (5.2.3)), χ^a must be timelike in a neighborhood of the horizon in regions I and II. Note, however, that χ^a need not be timelike everywhere in regions I and II. Without loss of generality, we may assume that χ^a is future-directed timelike in region I near the horizon.

Suppose, now, that χ^a is globally timelike in regions I and II. Then, we can perform the quantum field theory construction for a stationary spacetime (see section 4.3) in each of these two wedges separately. The essential features of this construction are the same as our "non-standard construction" of quantum field theory in

Minkowski spacetime based upon the boost Killing field b^a (see section 5.1). As in that construction, states in the Fock spaces obtained thereby have a natural interpretation in terms of particles "seen" by observers in regions I and II who follow orbits of χ^a.

In Minkowski spacetime, in addition to the one-parameter family of Lorentz boosts which generate the bifurcate Killing horizon, there exist the "ordinary time translation" isometries which are globally timelike over all of Minkowski spacetime and serve to define the standard quantum field theory construction and the usual Minkowski vacuum state $|0>_M$. However, in a general curved spacetime with bifurcate Killing horizon, there will not, in general, exist any globally timelike isometries which would enable us to perform a quantum field theory construction analogous to the standard construction for Minkowski spacetime. Nevertheless, we can proceed as follows. As discussed in section 5.1, in Minkowski spacetime, solutions in the one-particle Hilbert space, $\mathcal{H}_1$, associated with the standard construction are characterized by the fact that their restriction to h_A is purely positive frequency with respect to "affine time" V and, similarly, their restriction to h_B is purely positive frequency with respect to U. Now, suppose that in a curved spacetime with bifurcate horizon one had a vacuum state, ω—characterized, say, by the real inner product $\mu: \mathcal{S} \times \mathcal{S} \to \mathbb{R}$ or, equivalently, by the one-particle Hilbert space, $\mathcal{H}$, together with a real, linear map, $K: \mathcal{S} \to \mathcal{H}$, with dense range—having the property that for all $\psi \in \mathcal{S}$ the restriction of $K\psi$ to $h_A \cup h_B$ is nonvanishing and is purely positive frequency with respect to V on h_A and purely positive frequency with respect to U on h_B. Then an exact repetition of the analysis in Minkowski spacetime yields the following results: When ω is (formally) expressed as a state in $\mathcal{F}_s(\mathcal{H}_I) \otimes \mathcal{F}_s(\mathcal{H}_{II})$, it is precisely of the form (5.1.27) with *a* replaced by κ. Furthermore, the restriction of ω to region I is a thermal equilibrium state (with respect to the notion of time translations determined by χ^a) at the *Hawking temperature*, T, given by

$$T = \frac{\kappa}{2\pi} \tag{5.3.2}$$

Note that the temperature measured locally by any observer following an orbit of χ^a will differ from eq. (5.3.2) by the ratio of "Killing time" to "proper time" for that observer, so that the locally measured temperature is given by

$$T = \frac{\kappa}{2\pi\chi} \tag{5.3.3}$$

where χ is the "redshift factor", i.e., $\chi = (-\chi^a\chi_a)^{1/2}$. In particular, in view of eq. (5.2.6), this means that the temperature locally measured by a stationary observer very near the horizon will be given by the same formula, eq. (5.1.29), which holds for the Unruh effect in flat spacetime. This result could have been expected on the grounds that the state considered in the above argument corresponds closely to the Minkowski vacuum state, and one would not expect the (finite) spacetime curvature near the Killing horizon to have an important effect in the limit of arbitrarily large acceleration near the horizon.

The above discussion merely postulated the existence of a vacuum state, ω, which satisfies the property that $K\psi$ is purely positive frequency with respect to affine parameter on $\mathfrak{h}_A$ and $\mathfrak{h}_B$. However, there is strong motivation for considering states which satisfy this positive frequency condition. As discussed in detail in section 4.6, the expected stress-energy tensor of the quantum field in a state ω is well defined and nonsingular only when its two-point distribution, $\langle\phi(x_1)\phi(x_2)\rangle_\omega$, is of "Hadamard form" (see eq. (4.6.8)). It turns out that the Hadamard form of the two-point function of a vacuum state together with its invariance under the isometries generated by χ^a requires solutions in $\mathcal{H}$ to be purely positive frequency with respect to affine parameter on $\mathfrak{h}_A$ and $\mathfrak{h}_B$ in the following sense: Consider a Cauchy surface, Σ_A, containing a portion of $\mathfrak{h}_A$ as depicted in fig. 5.2 and consider a solution ψ_A whose restriction to Σ_A has support on $\mathfrak{h}_A$. (As explained in the "note in proof" in Kay and Wald (1991), such solutions, ψ_A, are not, in general, smooth [even if their restriction to Σ_A is smooth] and hence do not lie in $\mathcal{S}$. However, states satisfying the Hadamard condition are sufficiently regular that their action can be extended to an enlargement of Weyl algebra containing elements associated with a wide class of such ψ_A's.) Then, as shown by the detailed calculations given in Appendix B of Kay and Wald (1991), for any Hadamard vacuum state, ω, which is invariant under the isometries generated by χ^a, it follows that $\mu(\psi_A,\psi_A)$ must be equal to the Klein-Gordon norm of the positive frequency part of ψ_A on $\mathfrak{h}_A$ with respect to affine parameter V. Similar results hold for the analogous solutions, ψ_B, associated with the horizon, $\mathfrak{h}_B$. Thus, the condition that solutions in $\mathcal{H}$ be purely positive frequency with respect to affine parameter on $\mathfrak{h}_A$ and $\mathfrak{h}_B$

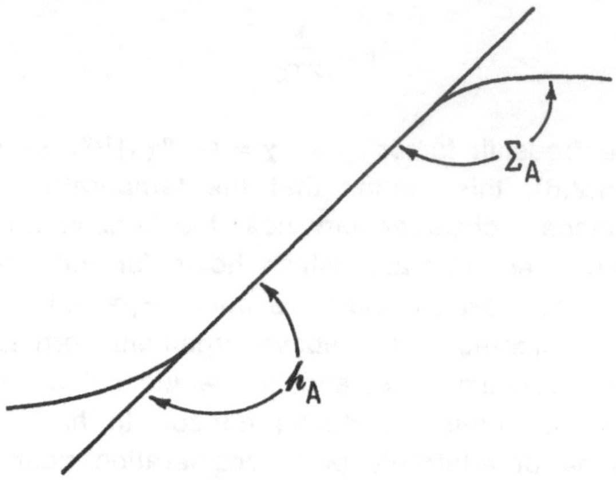

Figure 5.2. A spacetime diagram showing the Cauchy surface Σ_A.

actually is a consequence of nonsingularity (i.e., Hadamard form) of the vacuum state and its invariance under isometries generated by χ^a. In this regard, it should be noted that the Rindler vacuum state $|0>_I \otimes |0>_{II}$ in Minkowski spacetime (or its curved spacetime analogs, such as the Boulware vacuum of extended Schwarzschild spacetime) is not of Hadamard form on the horizon, and, indeed, its stress-energy tensor is divergent in the limit as the horizon is approached.

A mathematically precise and strengthened version of the above results may be stated as follows: Let $\mathcal{A}_0$ denote the subalgebra of the (enlarged) Weyl algebra generated by solutions of the form $\psi_A + \psi_B$. Let $\mathcal{A}_{0I}$ denote the subalgebra of $\mathcal{A}_0$ generated by solutions of the form $\psi_{AI} + \psi_{BI}$, where ψ_{AI} and ψ_{BI} vanish in region II. Let ω be any (not necessarily vacuum or even quasi-free) state which satisfies the Hadamard condition and is invariant under the isometries generated by χ^a. Then the restriction of ω to $\mathcal{A}_0$ is a (pure) vacuum state, and its restriction to $\mathcal{A}_{0I}$ is a thermal (i.e., KMS) state at the Hawking temperature (5.3.2). Furthermore, under an additional hypothesis (concerning the absence of "zero modes") ω is unique (if it exists) on the entire (enlarged) Weyl algebra. A complete proof of these results when ω is assumed to be quasi-free is given in Kay and Wald (1991). The strengthening of the result to drop the quasi-free assumption is given in Kay (1993).

The above results generalize the Unruh effect to curved spacetimes with a bifurcate Killing horizon. Indeed, these results strengthen the results of section 5.1 in that they show that the thermality of a state in regions I and II of Minkowski spacetime follows merely from its nonsingularity and its isometry invariance under b^a; the myriad of other special properties of the Minkowski vacuum state $|0>_M$ (such as invariance under the full Poincaré group) are not essential for its thermal nature with respect to b^a.

However, note that the curved spacetime results obtained above are weaker than the results we obtained for Minkowski spacetime in the following two ways: First, thermality is proven only on the ("large") subalgebra $\mathcal{A}_{0I}$ rather than on the full Weyl algebra for region I. Second, while uniqueness results are obtained, existence is not proven. Indeed, it is not difficult to show that if χ^a fails to be globally timelike throughout regions I and II—as occurs, for example, in Kerr spacetime—then no nonsingular, isometry invariant state, ω, can exist (see Kay and Wald (1991)). However, if χ^a is globally timelike in regions I and II and M is the union of the four "wedges" defined by the bifurcate Killing horizon (so that, in particular, there are no additional horizons), it seems plausible that existence will generally hold. In particular, as will be explained further below, if χ^a is hypersurface orthogonal (i.e., in the static case) and M is the union of the four "wedges", there is good reason to expect existence to hold unless "infra-red divergences" occur.

An interesting application of the generalization of the Unruh effect to curved spacetime is provided by deSitter spacetime. Suppose we have a nonsingular (i.e., Hadamard) state ω which is invariant under the full deSitter group. (As shown by Allen (1985), no such state exists for the massless Klein-Gordon field, but such a state was explicitly constructed by Allen in the massive case.) Then ω is a thermal state (in the appropriate "wedges" of deSitter spacetime) at temperature (5.3.2) for each Killing field in deSitter spacetime which generates a bifurcate Killing horizon. Now, given any timelike geodesic, γ, in deSitter spacetime, there exists a Killing field χ^a which is everywhere tangent to γ and which generates a bifurcate Killing horizon. We normalize χ^a so that $\chi^a\chi_a = -1$ on γ. Then the surface gravity of χ^a is readily computed to be

$$\kappa = \sqrt{\Lambda/3} \tag{5.3.4}$$

where Λ is the cosmological constant. Thus, the result (5.3.2) implies that for the nonsingular deSitter invariant vacuum state, ω, *every* inertial observer "feels himself to be immersed" in a thermal bath (and located in the "rest frame" of the bath) at temperature

$$T = \frac{1}{2\pi}\sqrt{\Lambda/3} \tag{5.3.5}$$

(Gibbons and Hawking 1977a).

Another good illustration of the curved spacetime Unruh effect is provided by extended Schwarzschild spacetime, depicted in fig. 5.3. The static Killing field χ^a of extended Schwarzschild spacetime generates a bifurcate Killing horizon. If we normalize χ^a so that $\chi^a\chi_a = -1$ at infinity, the surface gravity of χ^a is calculated to be

$$\kappa = \frac{1}{4M} \tag{5.3.6}$$

where M is the mass of the Schwarzschild spacetime. The analog in extended Schwarzschild spacetime of the Rindler vacuum state $|0\rangle_{\rm I} \otimes |0\rangle_{\rm II}$ (for which static observers detect no particles) is known as the *Boulware vacuum*. As with the Rindler vacuum, the expected stress-energy tensor in the Boulware vacuum is singular on both $\mathfrak{h}_A$ and $\mathfrak{h}_B$. The nonsingular vacuum state on extended Schwarzschild spacetime which is invariant under the isometries generated by χ^a is known as the Hartle-Hawking vacuum (Hartle and Hawking 1976; Israel 1976). By our results above, such a state will be a thermal state with respect to the notion of time translations generated by χ^a at temperature

$$T = \frac{1}{8\pi M} = \frac{\hbar c^3}{8\pi k G M} = 6 \times 10^{-8}\left(\frac{M_\odot}{M}\right) {}^\circ K \tag{5.3.7}$$

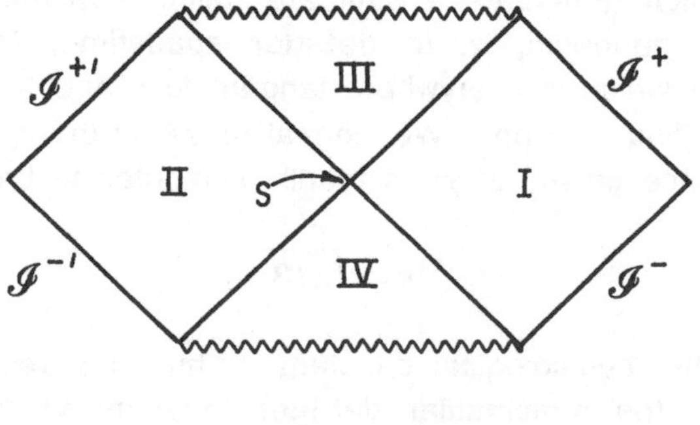

Figure 5.3. A conformal diagram of extended Schwarzschild spacetime.

Note that on account of our normalization of χ^a, eq. (5.3.7) corresponds to the temperature measured locally by static observers near infinity.

This thermal property of the Hartle-Hawking vacuum state often is presented as though it were a derivation of the effect of particle creation by a Schwarzschild black hole. This viewpoint is not correct. In a spacetime describing physical gravitational collapse to a black hole, only a portion of regions I and III of fig. 5.3 will be present; regions II and IV will be entirely absent. Hence, in a spacetime describing gravitational collapse, there is no motivation to impose a condition corresponding to nonsingularity of the state on the "past horizon" $\mathfrak{h}_B$. Indeed, when considered from the point of view of how it might be produced, the Hartle-Hawking vacuum state is highly unphysical, since it corresponds to having incoming thermal radiation from infinity in region I which is correlated with similar incoming radiation from infinity in region II, as well as with radiation emerging from the "white hole" singularity. In fact, as we shall see in chapter 7, the state corresponding to particle creation by a black hole at asymptotically late times after its formation by collapse is a thermal state at temperature (5.3.7) for modes which appear to emerge from the direction of the black hole, but is a vacuum state with respect to modes which are incoming from infinity. In extended Schwarzschild spacetime, this state corresponds to the "Unruh vacuum", which is singular on $\mathfrak{h}_B$.

The difference in nature between the "Unruh effect" we have been discussing here and the "Hawking effect" of particle creation by black holes (to be discussed in chapter 7) is dramatically illustrated by considering the case of the Kerr metric. As already noted above, it has been proven (Kay and Wald, 1991) that no analog of the Hartle-Hawking vacuum state exists in Kerr spacetime. Thus, there is no analog of the "Unruh effect" in Kerr spacetime. In essence, this is because there is no analog of incoming thermal radiation from infinity with respect to the notion of time translations defined by the Killing field which generates the horizon, since this Killing field has spacelike orbits near infinity. However, there is no corresponding difficulty with the derivation of particle creation in the case where gravitational collapse produces a Kerr black hole.

We conclude this section by giving an alternative "Euclidean derivation" of the Unruh effect which gives considerable insight into why the existence of a nonsingular thermal state at the temperature

(5.3.2) should be expected. However, this derivation is applicable only in the static case, i.e., when χ^a is hypersurface orthogonal.

First, we note that if a choice of μ has been made satisfying eq. (4.2.7) (or, more generally, eq. (4.5.11))—and a vacuum (or quasi-free) state $|0\rangle$ thereby has been defined—the Feynman propagator $\Delta_F: \mathfrak{I} \times \mathfrak{I} \to \mathbb{C}$ may be defined as the bi-distribution,

$$i\Delta_F(f,g) = \mu(Ef,Eg) + \frac{i}{2}\,[A(f,g) + R(f,g)] \qquad (5.3.8)$$

where, as before, A denotes the advanced Green's function, R denotes the retarded Green's function, and E = A−R (see eq. (3.2.22)). It can be seen from eq. (3.2.32) that eq. (5.3.8) corresponds to the formula

$$i\Delta_F(x,x') = \langle 0 \,|\, T[\hat{\phi}(x)\hat{\phi}(x')] \,|\, 0\rangle \qquad (5.3.9)$$

where T denotes the "time ordered product." It also follows immediately from eq. (5.3.8) that $\Delta_F(x,x')$ is a Green's function for the Klein-Gordon equation, which is symmetric in x and x' and whose real part is (A+R)/2.

Conversely, given any symmetric Green's function whose real part is (A+R)/2, we may reverse the above procedure and use eq. (5.3.8) to define μ. If this μ satisfies eq. (4.2.7) (or more generally, eq. (4.5.11)), the vacuum (or, respectively, quasi-free) state defined by μ will give rise to Δ_F via eq. (5.3.8)—or, equivalently, eq. (5.3.9)—as its associated Feynman propagator. Thus, the specification of a vacuum (or, more generally, quasi-free) state is equivalent to the specification of a suitable Green's function to serve as a Feynman propagator.

In the case of static spacetimes, the following "Euclidean procedure" is available for constructing a Feynman propagator: Let t denote the Killing parameter in the Lorentzian spacetime such that the level surfaces of t are orthogonal to the static Killing field χ^a. Then the formal substitution $t \to it$ will produce a Riemannian metric. The Riemannian space, M', thereby obtained together with the original Lorentzian spacetime, M, may then be viewed as different "real sections" of a single four-complex-dimensional manifold with complex metric. Now, if the Riemannian space is complete, the Klein-Gordon operator (which is elliptic in the Riemannian space) will be essentially self-adjoint on the Hilbert space $L^2(M')$ and therefore—in the absence of "zero modes"—a unique inverse can be defined. One thereby obtains a unique Green's function in the Riemannian space. One may then attempt to define a Feynman propagator in the Lorentzian spacetime by analytic continuation of this Green's function. In the case where χ^a is globally timelike, it can be shown (Wald 1979b) that this Euclidean construction of a Feynman propagator can indeed be carried out, and it coincides with the Feynman propagator obtained from the static vacuum state constructed by the procedure described in section 4.3.

In the case of interest here—where there is a bifurcate Killing horizon associated with χ^a and thus, in particular, χ^a fails to be globally timelike on M—the following interesting phenomenon occurs: The bifurcation surface, S, of the Lorentzian spacetime also lies on the "Euclidean section", so the analytically continued Killing field χ^a on the Riemannian manifold M' has the property of vanishing on the two-dimensional surface, S. As discussed in section 5.2, this corresponds to S being a "rotation axis" of the isometries generated by χ^a on M'. Hence, the only way for the Riemannian space M' to be nonsingular in a neighborhood of S is for the orbits of χ^a on M' to be periodic, with period, P, given by

$$P = 2\pi/\kappa \tag{5.3.10}$$

In other words, one will have a "conical singularity" at S unless such a periodic identification of the "imaginary time coordinate" is made.

If the Riemannian space, M', thereby obtained is complete, a unique Green's function on M' again can be defined. This Green's function obviously will have the periodicity (5.3.10) in "imaginary time". I am not aware of any theorems guaranteeing that such a Green's function on M' can be analytically continued to M so as to yield a Feynman propagator associated with a nonsingular (i.e., Hadamard) vacuum (or quasi-free) state, but it seems plausible that this will be the case except in circumstances where "infra-red divergences" occur (such as for a massless Klein-Gordon field in two-dimensional Minkowski spacetime). If so, then the resulting Feynman propagator clearly will have the property of being "periodic in imaginary time", with period given by eq. (5.3.10).

Now, it is well known that the property of a Feynman propagator being periodic in imaginary time is characteristic of a thermal state at temperature T = 1/P (see Fulling and Ruijsenaars (1987)).

Thus, assuming that the "Euclidean prescription" does define a nonsingular vacuum state, it will be a thermal state (with respect to the notion of time translations generated by χ^a) at temperature (5.3.2). Note that this "derivation" of the Unruh effect has a weaker status than the derivation previously described above in that a rigorous proof of existence of a nonsingular state has not been given (even on a subalgebra of the Weyl algebra). Furthermore, the properties of the state which make it unique (if it exists) concern the Riemannian Green's function rather than properties of the state in the physical Lorentzian spacetime. In addition, as noted above, this derivation is applicable only in the static case; if χ^a is not hypersurface orthogonal, analytic continuation in Killing parameter time t would not produce a real Riemannian metric. Nevertheless, this derivation relates in a very simple and clear way the existence of a thermal state at temperature (5.3.2) with geometrical properties of the (analytically continued) spacetime. Indeed, the basic argument does not depend upon the linearity of the quantum field (Gibbons and Perry 1978). For the case of a nonlinear field, it seems reasonable to presume that a nonsingular state could be defined on the Lorentzian spacetime by analytic continuation of n-point correlation functions defined on the Riemannian section. Such a state would then be a thermal equilibrium state at temperature (5.3.2). Thus, this derivation strongly suggests that the Unruh effect in curved spacetime will continue to hold for the case of nonlinear fields.

6 Classical Black Hole Thermodynamics

One of the most remarkable developments to take place in theoretical physics in the past few decades was the discovery of a close—and, undoubtedly, deep—relationship between laws of black hole physics and the ordinary laws of thermodynamics. In this chapter, we shall review the key aspects of this relationship within the context of classical general relativity. This will lay the foundation for our discussion, given in the next chapter, of the Hawking effect and its ramifications.

6.1 Classical Black Holes and the Area Theorem

In physical terms, a black hole is a region of spacetime where gravity is so strong that nothing—not even light—can escape. To define this notion more precisely, we must specify the region of spacetime to which possible "escape" is contemplated. We do so by restricting attention to spacetimes, (M, g_{ab}), which are asymptotically flat—i.e., (M, g_{ab}) becomes nearly Minkowskian "for all time" at "large distances" from some "central region"—and consider whether escape to this asymptotic region is possible. The precise notion of asymptotic flatness which is most useful for for formulating a definition of a black hole is that of "asymptotic flatness at null infinity": A spacetime, (M, g_{ab}), is said to be asymptotically flat at null infinity if we can conformally map it into another spacetime $(\tilde{M}, \tilde{g}_{ab})$ such that the image of M has null boundaries, $\mathcal{I}^+$ and $\mathcal{I}^-$, satisfying certain properties. The details of this definition can be found, e.g., in Wald (1984a). However, these details do not play a critical role in the discussion below, so the reader who is unfamiliar with this notion may substitute the vague phrase "asymptotically large distances at asymptotically late times" for "$\mathcal{I}^+$."

Consider, now, the family of all observers in the physical spacetime (M, g_{ab}) which escape to arbitrarily large distances at late times, i.e., more precisely, consider the collection of all timelike curves in M which have endpoints on $\mathcal{I}^+$ in $\tilde{M}$. If the past of these observers is not the entire spacetime, then a *black hole* is said to be present. This black hole, $\mathfrak{B}$, is comprised by all events in M which do not lie in the chronological past of of these observers; equivalently, we have

$$\mathfrak{B} = M - I^-(\mathscr{I}^+) \tag{6.1.1}$$

i.e., $\mathfrak{B}$ is the complement (in M) of the past (in $\tilde{M}$) of $\mathscr{I}^+$. The boundary of $\mathfrak{B}$ in M is called the (future) *event horizon* of the black hole. Equivalently, the event horizon is the boundary (within M) of the past (in $\tilde{M}$) of $\mathscr{I}^+$.

The notion of the event horizon of a black hole may be viewed as a specialization of the notion of the horizon associated with any observer or family of observers. Let (M, g_{ab}) be an arbitrary spacetime which is time oriented, i.e., a continuous choice of "future" vs. "past" on each lightcone can be (and has been) made. Let γ be an inextendible timelike curve, i.e., γ represents the world line of an observer in (M, g_{ab}). We define the *future horizon*, h^+, of γ to be the boundary, $\dot{I}^-(\gamma)$, of the chronological past, $I^-(\gamma)$, of γ. The past horizon, h^-, of γ is defined similarly. The definitions of h^+ and h^- can be straightforwardly extended to families of observers: For a family of timelike curves $\{\gamma_\alpha\}$, h^+ is defined to be the boundary of the past of the union of the γ_α.

Note that the notion of the future or past horizon of an observer or family of observers is entirely independent of the notion of a Killing horizon introduced in section 5.2 above. However, there are some important relationships between these logically distinct notions: First, in any spacetime with a bifurcate Killing horizon, let $\{\gamma_\alpha\}$ denote the family of Killing orbits in region I (see eq. (5.3.1a)) which are timelike. Then the future horizon, h^+, of these observers includes (and possibly coincides with) $\mathfrak{h}_A$, and the past horizon, h^-, includes $\mathfrak{h}_B$. In particular, in Minkowski spacetime, the family $\{\gamma_\alpha\}$ is comprised by uniformly accelerating observers. In this case, the future and past horizons, h^+ and h^-, of any one of these observers coincide with the portions, $\mathfrak{h}_A$ and $\mathfrak{h}_B$, respectively, of the bifurcate Killing horizon shown in fig. 5.1 above. A second important relationship between the notions of future and past horizons and Killing horizons was already mentioned in section 5.2 and will be discussed further in the next section: The event horizon of a stationary black hole must be a Killing horizon.

Some crucial properties of h^+ follow directly from its definition. We summarize them in the following theorem:

Theorem 6.1.1 : The future horizon, h^+, of any observer or family of observers, is a (C^0) three-dimensional, achronal submanifold. Each $p \in h^+$ lies on a future inextendible null geodesic segment, α, which is contained entirely in h^+. Furthermore, the convergence (defined below) of these null geodesics generating h^+ cannot become infinite at any point on h^+.

The first two of these properties of h^+ are direct consequences of standard theorems on causal structure (see, e.g., theorems 8.1.3 and 8.1.6 of Wald (1984a)). The possibility—left open by these theorems—that α could have a future endpoint on γ rather than be inextendible, is ruled out by the fact that $\gamma \subset I^-(\gamma)$. The last property of theorem 6.1.1 follows from the fact that the existence of a point $q \in h^+$ at which the convergence of the null generators is infinite would violate the achronality of h^+ (see theorem 9.3.10 of Wald (1984a) and the related discussion).

The definition we have just given of a black hole permits the occurrence of "naked singularities", i.e., singularities which form outside of the event horizon and thereby could be visible to—and have influence upon—observers at large distances from the black hole. If naked singularities could be present, very little could be said about the event horizon of a black hole beyond the general results contained in theorem 6.1.1. However, it is widely believed that no such naked singularities will occur in a physically realistic gravitational collapse. As we shall see shortly, if this is the case, then some key additional properties—in particular, the area theorem—will be satisfied by an event horizon.

The conjecture that "no naked singularities occur" is known as the "cosmic censor hypothesis", and can be formulated in a relatively precise manner as follows:

Cosmic Censor Hypothesis: Consider asymptotically flat initial data on a spacelike hypersurface (with compact "interior region") for a solution of Einstein's equation with "suitable" matter. Then the maximal Cauchy development of these data (i.e., the "largest" spacetime uniquely determined by these data and Einstein's equation) "almost always" is asymptotically flat at null infinity (with the generators of $\mathscr{I}^+$ being complete).

The terms "suitable" and "almost always" appearing in this formulation of the cosmic censor hypothesis require some further explanation. Two obvious necessary conditions on matter for it to be "suitable" are that it be governed by deterministic (i.e., hyperbolic) differential equations and that it have locally positive energy density (more precisely, that its stress-energy tensor obey the dominant energy condition). However, perfect fluids with certain equations of state (such as "dust") obey both of these conditions, and yet are known to yield solutions with naked singularities. The naked singularities occurring in these solutions are generally viewed as resulting from the fluid matter not being a "suitable" model for the fundamental matter fields occurring in nature, rather than as an indication of the failure of cosmic censorship. A possible additional requirement on matter fields for them to be "suitable" is that when their differential equations are evolved on a fixed, nonsingular, globally hyperbolic spacetime (such as Minkowski spacetime), one always obtains globally nonsingular solutions. Consequently, any singularities occurring in the Einstein-matter system necessarily would be attributable to gravitational effects. The types of perfect fluid matter which are known to yield examples with naked singularites do not satisfy this additional requirement.

The "almost always" condition was inserted in the above formulation because it would not be fatal to the physical content of the conjecture if some counterexamples exist, provided that the initial data required for these counterexamples is so special (i.e., a "set of measure zero") that it would be physically impossible to achieve. Indeed, the types of fluid counterexamples mentioned above may be of this character, so even if one wished to include fluids as "suitable" matter, it is possible that no physically achieveable counterexamples could be constructed with fluids. The examples of spherically symmetric Einstein-scalar-field solutions with naked singularities recently found by Christodoulou (1994) are known to be of a "non-generic" character.

In the more than twenty years since the cosmic censor conjecture was first posed, little progress has been made toward obtaining a direct, general proof of it. However, most (but not all) researchers have been convinced of its validity by some partial results, the failure of some proposed counterexamples, and the remarkable consistency of the picture of gravitational collapse to which it leads. A good example of this internal consistency can be found in our discussion of the "first law" given in section 6.2 below. Further discussion of the evidence both for and against the validity of the cosmic censor hypothesis can be found in Wald (1993b).

Let (M, g_{ab}) be an asymptotically flat solution of Einstein's equation with "suitable" matter, which contains an asymptotically flat slice Σ with compact "interior region", and is such that M contains the maximal Cauchy development of data on Σ. Then, if the cosmic censor hypothesis is valid, generically, M will be asymptotically flat at $\mathscr{I}^+$, and the domain of dependence, D(Σ), of Σ in M (see eq. (4.1.3)) will include all events to the future of Σ which are "visible" from infinity i.e., it will include $I^+(\Sigma) \cap I^-(\mathscr{I}^+)$. This means that, generically, phenomena in the entire region exterior to any black hole, $\mathcal{B}$, that may form is "predictable" from Σ in the sense of theorem 4.1.2. It does not follow (even if the above formulation of cosmic censorship holds) that any events in $\mathcal{B}$ itself need be contained in D(Σ). However, the requirement that the event horizon, h^+, of $\mathcal{B}$ be contained in D(Σ) would involve only a modest strengthening of the above formulation of the cosmic censor hypothesis. If, in addition to $[I^+(\Sigma) \cap I^-(\mathscr{I}^+)] \subset D(\Sigma)$, we have $h^+ \subset D(\Sigma)$, then $\mathcal{B}$ is said to be *predictable*. In other words, an asymptotically flat spacetime (M, g_{ab}) is said to contain a predictable black hole if there exists a globally hyperbolic region $\mathcal{O} \subset M$ which contains both the region exterior to the black hole and the event horizon of the black hole.

The area theorem holds for the event horizon of a predictable black hole. In order to prove it, we need to develop some simple properties of a family of null geodesics which generate a null hypersurface, $\mathcal{N}$, such as an event horizon. Let λ denote an affine parametrization of the null geodesic generators of $\mathcal{N}$, and let k^a denote their tangents with respect to this parametrization. Let α be a null geodesic generator of $\mathcal{N}$, and let $p \in \alpha$. The expansion, θ, of the null geodesic generators of $\mathcal{N}$ at a point p is defined by $\theta = \nabla_a k^a$. Equivalently, consider an infinitesimal cross-sectional area element of area A at p and carry this area element (i.e., Lie transport it) along the null geodesic generators of $\mathcal{N}$. Then θ is given by

$$\theta = \frac{1}{A}\frac{dA}{d\lambda} \tag{6.1.2}$$

i.e., θ measures the local rate of change of cross-sectional area as one moves up the geodesics.

The geodesic deviation equation governs the rate of change of θ. Using it, we obtain the Raychauduri equation (see, e.g., Wald (1984a)),

$$\frac{d\theta}{d\lambda} = -\frac{1}{2}\theta^2 - \sigma_{ab}\sigma^{ab} - R_{ab}k^a k^b \tag{6.1.3}$$

where σ_{ab} denotes the shear of the geodesics and R_{ab} is the Ricci curvature of the spacetime. Now, if Einstein's equation holds with the stress-energy tensor T_{ab} satisfying the null energy condition $T_{ab}k^a k^b \geqq 0$, we obtain $R_{ab}k^a k^b \geqq 0$ and, hence,

$$\frac{d\theta}{d\lambda} \leqq -\frac{1}{2}\theta^2 \tag{6.1.4}$$

from which it immediately follows that

$$\theta^{-1}(\lambda) \geqq \theta_0^{-1} + \frac{1}{2}\lambda \tag{6.1.5}$$

where θ_0 denotes the initial value of θ. Thus, if $\theta_0 < 0$ (i.e., the geodesics initially are converging) we find that $\theta(\lambda_1) = -\infty$ (i.e., the convergence is infinite) at some $\lambda_1 \leqq 2/|\theta_0|$ provided, of course, that the geodesic α can be extended that far.

The machinery now is in place to state and prove the area theorem:

Theorem 6.1.2 (Area theorem): For a predictable black hole satisfying $R_{ab}k^a k^b \geqq 0$ for all null k^a, the surface area of the future event horizon, h^+, never decreases with time.

Proof : We show, first, that we have $\theta \geqq 0$ everywhere on the event horizon h^+ of a predictable black hole, $\mathcal{B}$. If the null generators of the horizon were assumed to be geodesically complete (as in the original proof of Hawking (1971)), this conclusion actually would follow immediately from the above result together with the fact that, by theorem 6.1.1 above, we cannot have $\theta = -\infty$ at any point of h^+. However, for a predictable black hole, it is not necessary to assume completeness of the generators of h^+ (Hawking and Ellis 1973): If $\theta < 0$ at $p \in h^+$, we could deform a two-dimensional cross-section, S, of the horizon outward in a small neighborhood of p, keeping the expansion of the orthogonal null geodesics negative. Let S' denote this deformed cross-section. Then $\mathscr{I}^+ \cap I^+(S')$ is nonempty, and global hyperbolicity implies that any point $q \in \mathscr{I}^+ \cap \dot{I}^+(S')$ must be connected to S' by a null geodesic, α, which lies in $\dot{I}^+(S')$ (see theorem 8.3.11 of Wald (1984a)). Furthermore, α is orthogonal to S' and is future complete, so that $\theta = -\infty$ at some point of α for the congruence of null geodesics orthogonal to S'. Application of theorem 6.1.1 to the "deformed horizon"—i.e., the horizon associated with the family of observers who have endpoints on $\mathscr{I}^+ \cap \dot{I}^+(S')$—then yields a contradiction. Note that this argument requires global hyperbolicity of only the region exterior to the black hole. Note also that the same argument shows that if the region exterior to a black hole is globally hyperbolic, then any trapped surface must be entirely contained within the black hole. (Here, a ***trapped surface*** is a compact, two-dimensional, spacelike submanifold having the property that both sets of orthogonal null geodesics (i.e., "ingoing" and "outgoing") have negative expansion, $\theta < 0$.)

Since $\theta \geqq 0$, the cross-sectional area of the horizon locally increases as one moves up the generators of h^+. By theorem 6.1.1, these generators cannot leave h^+ (although new generators can join h^+). Nevertheless, one has to worry about the possibility that these null geodesic generators might not reach a sufficiently late time slice Σ (e.g., they might terminate on a singularity on the horizon), thus causing the area of $h^+ \cap \Sigma$ to be smaller than the initial area. However, this possibility cannot occur for a predictable black hole, wherein the event horizon as well as the exterior region is contained in a globally hyperbolic region, $\mathcal{O}$, of M. Namely, if Σ is a Cauchy surface for this region, then by a standard theorem (see theorem 8.3.7 of Wald (1984a)) every null geodesic in $\mathcal{O}$ must intersect Σ. Thus, if Σ_1 and Σ_2 are Cauchy surfaces with $\Sigma_2 \subset I^+(\Sigma_1)$, every generator of h^+ at Σ_1 must reach Σ_2. Thus the area of $h^+ \cap \Sigma_2$ must be at least as large as the area of $h^+ \cap \Sigma_1$, as we desired to show. □

6.2 The Laws of Black Hole Mechanics

The statement of the area theorem bears a superficial resemblence to the statement of the second law of thermodynamics in that both results assert that a certain quantity never decreases with time. In this section, we shall show that other laws of black hole mechanics bear a striking mathematical similarity to the laws of thermodynamics.

We begin by recalling the following result due to Hawking (see proposition 9.3.6 of Hawking and Ellis (1973)): Let (M, g_{ab}) be an asymptotically flat spacetime which is stationary, i.e., there exists a one-parameter group of isometries whose Killing field ξ^a is timelike near infinity. Suppose, further, that (M, g_{ab}) is a solution of Einstein's equation with matter satisfying suitable hyperbolic equations, and that the metric and matter fields are analytic. (Analyticity of the metric and matter fields in the region where ξ^a is timelike follows as a consequence of the ellipticity of the stationary field equations, but it is an additional assumption elsewhere.) Then the event horizon, h^+, of any black hole, $\mathcal{B}$, in (M, g_{ab}) is a Killing horizon, i.e., there exists a Killing field χ^a (not necessarily coinciding with ξ^a) which is normal to h^+.

Since h^+ must be invariant (i.e., mapped into itself) under the stationary isometries, it is obvious that the stationary Killing field ξ^a must be tangent to h^+. The above result states that if ξ^a fails to be normal to h^+, then there exists an additional Killing field, χ^a, which is. Furthermore, in this case where $\chi^a \neq \xi^a$, it can be shown that a linear combination, φ^a, of these two Killing fields, can be chosen so that the orbits of φ^a are closed. Thus, if $\chi^a \neq \xi^a$, then (M, g_{ab}) is axisymmetric as well as stationary. For a stationary black hole, we define the *angular velocity of the horizon*, Ω, by,

$$\chi^a = \xi^a + \Omega\, \varphi^a \tag{6.2.1}$$

where φ^a is normalized by requiring that the closed orbits have period 2π, and ξ^a is normalized by requiring $\xi^a\xi_a \to -1$ at infinity.

The above result does not put any restriction on the type of Killing horizon that may occur, e.g., it does not imply that the event horizon of a stationary black hole must comprise a portion of a bifurcate Killing horizon (see section 5.2). However, it can be shown by a rather lengthy calculation (Bardeen et al. 1973) that if Einstein's equation holds with matter satisfying the dominant energy condition, then the surface gravity, κ, (defined by eq. (5.2.4) above) of any (connected) Killing horizon, h, must be constant over the horizon. However, if κ is constant and nonzero over h, then it can be shown (Racz and Wald 1992) that one always can locally extend, if necessary, a neighborhood of h so that h comprises a portion of a bifurcate Killing horizon (although existence of a global extension with this property was not proven). This indicates that the only physically relevant types of Killing horizons—i.e., the only Killing horizons that occur in solutions of Einstein's equation with matter satisfying the dominant energy condition—are bifurcate Killing horizons and "degenerate horizons" (defined as those for which $\kappa = 0$). In particular, except in the degenerate case, the event horizon, h^+, of a stationary black hole should always correspond to the portion, $\mathfrak{h}_A$, of a bifurcate Killing horizon. This is indeed the case for all the known black hole solutions. In particular, the charged Kerr solutions with $Q^2 + (J/M)^2 \leq M^2$ comprise all the known stationary black hole solutions of the Einstein-Maxwell equations. The horizons of the charged Kerr black holes with $Q^2 + (J/M)^2 < M^2$ correspond to the bifurcate case (and have been proven to be the only solutions with $\kappa \neq 0$), whereas the horizons of the "extreme" charged Kerr black holes satisfying $Q^2 + (J/M)^2 = M^2$ correspond to the degenerate case.

We turn, now to the derivation of the "first law" of black hole mechanics. Actually, there are two logically independent versions of this law, which I will refer to as the "physical process version" and the "equilibrium state version."

For the "physical process version," we start with a stationary black hole and alter it by some (infinitesimal) physical process. We assume that the black hole is not destroyed by this process and that it "settles down" to a new stationary final state. We then relate its change in area to the changes in the other parameters describing the black hole.

For simplicity, we treat the case of a vacuum black hole which is altered by dumping in a small amount of matter represented by stress energy ΔT_{ab}. Then, to first order in ΔT_{ab}, we can neglect the change in the black hole geometry when computing the resulting changes in mass, ΔM, and angular momentum, ΔJ, of the black hole. Thus, we obtain

$$\Delta M = \int_0^\infty dV \int d^2S \ \ \Delta T_{ab} \ \xi^a \ k^b \tag{6.2.2}$$

$$\Delta J = -\int_0^\infty dV \int d^2S \ \ \Delta T_{ab} \ \varphi^a \ k^b \tag{6.2.3}$$

where we have chosen to parametrize the null geodesics (with tangent k^a) on the horizon by affine parameter, V, and the second integral in these expressions is over the cross-section, S, of the horizon corresponding to the "time" V. [Note that the product $k^b dV$ is parametrization-independent. Note also that the range of integration from 0 to ∞ for V corresponds to the range $-\infty$ to $+\infty$ of Killing parameter v (see eq. (5.2.10)).] On the other hand, the change in area is governed by the Raychaudhuri equation (6.1.3) applied to the exact horizon. To first order in ΔT_{ab}, the quadratic terms θ^2 and $\sigma_{ab}\sigma^{ab}$ can be neglected. Hence, using Einstein's equation, we obtain,

$$\frac{d\theta}{dV} = -8\pi\, \Delta T_{ab}\, k^a k^b \tag{6.2.4}$$

When integrating the right side of this equation over the horizon, the change in the black hole geometry may be neglected. Thus, on the right side of eq. (6.2.4), we may substitute for k^a

$$\begin{aligned} k^a = (\partial/\partial V)^a &= \frac{1}{\kappa V}(\partial/\partial v)^a = \frac{1}{\kappa V}\chi^a \\ &= \frac{1}{\kappa V}(\xi^a + \Omega\varphi^a) \end{aligned} \tag{6.2.5}$$

Hence, multiplying both sides of eq. (6.2.4) by κV and integrating over the horizon, we obtain (Hawking and Hartle 1972)

$$\begin{aligned} \kappa \int_0^\infty dV \int d^2S\, V\frac{d\theta}{dV} &= -8\pi \int_0^\infty dV \int d^2S\, \Delta T_{ab}(\xi^a + \Omega\varphi^a)k^b \\ &= -8\pi\,(\Delta M - \Omega \Delta J) \end{aligned} \tag{6.2.6}$$

The left side of this equation can be evaluated by integration by parts,

$$\int d^2S \left(\int_0^\infty V\, \frac{d\theta}{dV}\, dV\right) = \int d^2S\, (\theta V)\, \Big|_0^\infty - \int d^2S \int_0^\infty \theta\, dV \tag{6.2.7}$$

By eq. (6.1.2), the second term on the right side is just minus the change in area of the black hole. On the other hand, the first term vanishes since $V = 0$ at the lower limit and θ must vanish faster than $1/V$ as $V \to \infty$ if the black hole settles down to a stationary final state with finite area. Thus, we obtain

$$\kappa\, \Delta A = 8\pi\, (\Delta M - \Omega\, \Delta J) \tag{6.2.8}$$

which is the desired "physical process" version of the first law. Note that eq. (6.2.8) together with the equality of the second and third terms in eq. (6.2.6) shows that $\Delta A \geqq 0$ provided that $\Delta T_{ab} k^a k^b \geqq 0$, in accordance with the area theorem.

The "equilibrium state" version of the first law simply compares the areas of two infinitesimally nearby stationary black hole solutions to Einstein's equation. The original derivation that eq.(6.2.8) holds for nearby stationary solutions involved a rather lengthy calculation (Bardeen et al. 1973). However, it turns out that eq. (6.2.8) can be derived in a very simple, direct, and general manner from the Hamiltonian or Lagrangian formulations of general relativity (Sudarsky and Wald 1992, Wald 1993a,c, Iyer and Wald 1994). We now shall outline the derivation of the equilibrium state version of the first law for general relativity using the Hamiltonian formulation, and then shall briefly sketch how the first law may be derived for an arbitrary diffeomorphism invariant theory of gravity obtained from a Lagrangian.

It is well known that general relativity can be given a Hamiltonian formulation (see e.g., Appendix E of Wald (1984a)). A point in the phase space of general relativity corresponds to the specification of the fields (h_{ij}, π^{ij}) on a three dimensional manifold Σ. Here h_{ij} is a Riemannian metric on Σ and its canonically conjugate momentum is $(1/16\pi)\pi^{ij}$ where π^{ij} is related to the extrinsic curvature, K_{ij}, of Σ in the spacetime obtained by evolving this initial data by

$$\pi^{ij} = \sqrt{h}\,(K^{ij} - h^{ij} K) \qquad (6.2.9)$$

(The letters "i,j,k,..." are used here for "spatial indices", i.e., abstract indices of tensors on Σ; as elsewhere in this book, the letters "a,b,c,..." are used for "spacetime indices", i.e., abstract indices of tensors on M.) As also is well known, constraints are present in general relativity. In the vacuum case, the allowed initial data are restricted to the constraint submanifold in phase space defined by the vanishing at each point $x \in \Sigma$ of the quantities,

$$0 = \mathcal{C}_0 = \frac{1}{16\pi}\sqrt{h}\left\{-{}^{(3)}R + \frac{1}{h}\left[\pi^{ij}\pi_{ij} - \frac{1}{2}\pi^2\right]\right\} \qquad (6.2.10)$$

$$0 = \mathcal{C}_i = -\frac{1}{8\pi}\sqrt{h}\; D_j(\pi_i{}^j/\sqrt{h}) \qquad (6.2.11)$$

where D_i denotes the derivative operator on Σ compatible with h_{ij} and ${}^{(3)}R$ denotes the scalar curvature of h_{ij}.

The ADM Hamiltonian, H, for general relativity has the "pure constraint" form,

$$H = \int_\Sigma N^a \mathcal{C}_a \tag{6.2.12}$$

where N^a is to be viewed as a nondynamical variable, which may be prescribed arbitrarily. In the spacetime obtained by solving Hamilton's equations, N^a has the interpretation of being the time evolution vector field, i.e., its projection normal to Σ yields the lapse function, N, and its projection into Σ yields the shift vector, N^i. This "pure constraint" form of H is not special to the vacuum Einstein equations; any diffeomorphism invariant theory which can be given a Hamiltonian formulation always has a Hamiltonian of this general form (see the Appendix of Lee and Wald (1990) and Iyer and Wald (1994)).

Now, let (h_{ij}, π^{ij}) be any initial data satisfying the constraints (6.2.10)-(6.2.11). Choose any time evolution vector field N^a on Σ. Let $(\delta h_{ij}, \delta\pi^{ij})$ be an arbitrary perturbation of these data (not necessarily satisfying the linearized constraints). Then the variation of eq. (6.2.12) yields an equation of the form

$$\delta H = \int_\Sigma \left[\mathcal{P}^{ij}\,\delta h_{ij} + \mathcal{Q}_{ij}\,\delta\pi^{ij}\right] + \text{"surface terms"} \tag{6.2.13}$$

where the "surface terms" arise from the integrations by parts needed to remove the spatial derivatives from δh_{ij} and $\delta\pi^{ij}$. The statement that H is a Hamiltonian for general relativity means precisely that the coefficient, $\mathcal{P}^{ij}$, of δh_{ij} equals minus the "time derivative" (i.e., the Lie derivative with respect to N^a) of the canonical momentum $(1/16\pi)\pi^{ij}$ in the solution to Einstein's equations arising from the initial data (h_{ij}, π^{ij}), and that the coefficient, $16\pi\mathcal{Q}_{ij}$, of $(1/16\pi)\delta\pi^{ij}$ equals the time derivative of h_{ab}. Explicit formulas for $\mathcal{P}^{ij}$ and $\mathcal{Q}_{ij}$ can be found, e.g., in Appendix E of Wald (1984a).

Now, consider a stationary black hole with bifurcate Killing horizon. Choose an asymptotically flat hypersurface, Σ, which intersects the bifurcation surface, σ, of the horizon, and let (h_{ij}, π^{ij}) be the initial data which are induced on Σ. Choose the Killing field χ^a, eq. (6.2.1), which is normal to the horizon (and thus vanishes at σ) as the time evolution vector field, N^a. Let $(\delta h_{ij}, \delta\pi^{ij})$ be any smooth perturbation of these data which is asymptotically flat and satisfies the linearized constraints (i.e., $\delta\mathcal{C}_a = 0$). Then it follows immediately from eq. (6.2.12) that for this perturbation, we have $\delta H = 0$. On the other hand, the fact that χ^a is a Killing field of the background solution implies that $\mathcal{P}^{ij} = \mathcal{Q}_{ij} = 0$. Consequently, eq. (6.2.13) yields the result that

$$\text{"surface terms"} = 0 \qquad (6.2.14)$$

These surface terms are readily evaluated. Since $N^a = \chi^a$ approaches a linear combination of a time translation and rotation at infinity (see eq. (6.2.1)), the surface term from infinity is directly related to the changes, ΔM and ΔJ, in the mass and angular momentum ("as measured at infinity") produced by the perturbation $(\delta h_{ij}, \delta\pi^{ij})$. On the other hand, the surface term from σ is readily evaluated to be proportional to κ times the change, ΔA, in the area of σ produced by the perturbation (see Sudarsky and Wald (1992)). Consequently, eq. (6.2.14) takes the explicit form

$$\frac{1}{8\pi}\kappa\,\Delta A = \Delta M - \Omega\,\Delta J \qquad (6.2.15)$$

which is our desired "equilibrium state" version of the first law. Note that our derivation shows that eq. (6.2.15) actually holds for any (nonsingular, asymptotically flat) perturbation satisfying the constraints, i.e., the perturbation may be nonstationary. Note also that this derivation admits a straight-forward generalization to the case where matter fields are present, provided only that the combined Einstein-matter system can be given a Hamiltonian formulation (see Sudarsky and Wald (1992) for the case of the Einstein-Yang-Mills system).

Indeed, much more generally, the basic form of eq. (6.2.15) can be shown to hold in an arbitrary diffeomorphism invariant theory of gravity (with arbitrary matter fields) derived from a Lagrangian (Wald 1993c, Iyer and Wald 1994). In this approach, one starts from a Lagrangian of the general form

$$L = L(g_{ab}, R_{abcd}, \nabla_e R_{abcd}, \ldots; \psi, \nabla_a\psi, \ldots) \qquad (6.2.16)$$

in an n-dimensional spacetime, where ψ denotes the collection of matter fields present in the theory. For each vector field N^a on spacetime, one defines a Noether current, (most conveniently viewed as an (n−1)-form, $\mathbf{j}$) associated with the local symmetry corresponding to the infinitesimal diffeomorphism generated by N^a. When the equations of motion hold, the Noether current can be derived from a potential, i.e., we have $\mathbf{j} = d\mathbf{Q}$ for an (n–2)-form, $\mathbf{Q}$, referred to as the *Noether charge*. Integration over the hypersurface Σ of the preceding paragraph of a general identity involving the variation of the Noether current then yields an equation of the general form (6.2.15), with the left side replaced by the variation, $\Delta\int_\sigma \mathbf{Q}$, of the Noether charge associated the Killing field χ^a integrated over the bifurcation surface σ (Wald, 1993c). Further analysis (Iyer and Wald 1994) shows that

$$\Delta\int_\sigma \mathbf{Q} = \frac{\kappa}{2\pi}\Delta S \tag{6.2.17}$$

where S is given by

$$S = -2\pi\int_\sigma E^{abcd}\, n_{ab} n_{cd} \tag{6.2.18}$$

where n_{ab} denotes the binormal to σ, and E^{abcd} is simply the functional derivative of L with respect to R_{abcd} holding g_{ab} (and ∇_a) fixed,

$$E^{abcd} \equiv \frac{\delta L}{\delta R_{abcd}} \tag{6.2.19}$$

i.e., $E^{abcd} = 0$ would be the Euler-Lagrange equation of motion for R_{abcd} if one pretended that it were a dynamical field, independent of the metric, in the Lagrangian (6.2.16). Thus, for example, in the case of vacuum general relativity, we have

$$L = \frac{1}{16\pi} R = \frac{1}{16\pi} g^{ac} g^{bd} R_{abcd} \tag{6.2.20}$$

so we simply obtain

$$E^{abcd} = \frac{1}{16\pi} g^{ac} g^{bd} \tag{6.2.21}$$

and

$$\begin{aligned} S &= -2\pi\int_\sigma \frac{1}{16\pi} g^{ac} g^{bd}\, n_{ab} n_{cd} \\ &= -\frac{1}{8}\int_\sigma n^{ab} n_{ab} = \frac{1}{4} A \end{aligned} \tag{6.2.22}$$

Hence, we recover eq. (6.2.15). However, this derivation shows that eq. (6.2.15) continues to hold in an arbitrary diffeomorphism invariant theory provided that we replace the quantity A/4 by the local, geometrical quantity (6.2.18).

We remark that comparison of eqs. (6.2.8) and (6.2.15) shows that the differences in mass, angular momentum, and area of two nearby stationary black hole solutions satisfy precisely the relationship required by the "physical process" argument given above. If the "physical process" and "equilibrium state" versions of the first law disagreed, it would establish an inconsistency in the assumptions that went into the derivation of the "physical process" version, particularly the assumptions that the black hole is not destroyed by the process of throwing matter into it and that it "settles down" to a stationary final state. The fact that the two versions of the first law do agree is correspondingly strong support for the consistency of the idea that gravitational collapse always results in a predictable black hole.

We are now in a position to describe the remarkable mathematical analogy between the ordinary laws of thermodynamics and certain laws (derived above) satisfied by black holes in classical general relativity. First we explain the framework of this analogy. We view a black hole as being analogous to an ordinary dynamical system with many degrees of freedom. Now, for a general, nonequilibrium state of an ordinary dynamical system, we must use the detailed, "microscopic" structure of the system to describe its state, and we must use the full "microscopic" dynamical laws to determine its time evolution. Similarly, for a general, nonstationary black hole, we must use the detailed initial data of general relativity to describe its state and the full Einstein equation to determine its time evolution. However, states of thermal equilibrium of an ordinary dynamical system can be characterized by a very small number of "state parameters" such as total energy, E, and volume, V. Similarly, stationary black hole states also can be characterized by a very small number of parameters. Indeed, it is known that in the electrovac case, the mass, M, angular momentum, J, and electric and magnetic charges, Q_E and Q_M, characterize a stationary black hole (see Wald (1984a) for a summary of the black hole uniqueness theorems). For thermal equilibrium states, one can define additional quantities such as temperature, T, and pressure, P. Similarly, for a stationary black hole—where the event horizon is a Killing horizon—

we can define κ and Ω as described above. Finally, there is an additional thermodynamic quantity, the total entropy, S, which also can be defined for nonequilibrium states. Similarly, the surface area, A, of the event horizon also is defined for nonstationary black holes.

The above analogy in the framework of black hole physics and the thermodynamics of ordinary dynamical systems is not particularly striking by itself. What is quite striking, however, is the analogy between the area theorem 6.1.2 and the ordinary second law of thermodynamics, which states that the total entropy, S, never decreases with time. Now, it might appear that the nature of the area theorem and the ordinary second law could hardly be more different: The area theorem is a rigorous theorem in differential geometry applicable to predictable black holes satisfying $R_{ab}k^a k^b \geqq 0$. The time asymmetry in the area theorem arises from the fact one is dealing with a future horizon, h^+, rather than a past horizon. On the other hand, the ordinary second law is not believed to be a rigorous law but rather one which holds with (overwhelmingly) high probability. The time asymmetry of this law arises from a choice of (highly improbable) initial state. Nevertheless, there are very few laws of physics which involve time asymmetric behavior of a quantity—allowing it to increase with time but not decrease—so the analogy between these laws should not be dismissed.

The analogy between laws of black hole physics and thermodynamics is strongly reinforced when one compares eq. (6.2.15) (or, equivalently, eq. (6.2.8)) with the ordinary first law of thermodynamics:

$$T\Delta S = \Delta E + P\Delta V \tag{6.2.23}$$

(It is worth noting that for a given thermodynamic system, one typically obtains both a "physical process" and "equilibrium state" version of this ordinary first law, although one commonly does not emphasize the fact that the consistency of these two versions supports the conjecture that all physically realistic initial states of the dynamical system asymptotically approach a thermal equilibrium state.) The term "$-\Omega\Delta J$" in eq. (6.2.15) is closely analogous to the "work term" $P\Delta V$ in eq. (6.2.23). Indeed, if one considers a rotating body in ordinary thermodynamics, one obtains precisely such a "$-\Omega\Delta J$" term in the ordinary first law. Thus, eqs. (6.2.15) and (6.2.23) have remarkably similar characters.

The analogy is further strengthened by consideration of the "zeroth law" which, in ordinary thermodynamics, states that for a body in thermal equilibrium, the temperature must be uniform over the body. Comparing eqs. (6.2.15) and (6.2.23), we see that the quantity in black hole physics which plays the role analogous to temperature, T, is the surface gravity, κ. As discussed above, for a stationary black hole κ must be uniform over the event horizon.

One formulation of the third law of ordinary thermodynamics asserts that $S \to 0$ (or a universal constant) as $T \to 0$. The analog of this law is not satisfied in black hole physics, since, in particular, $\kappa = 0$ for all "extreme" ($J = M^2$) Kerr black holes, which have variable $A \neq 0$. However, while this formulation of the ordinary third law is believed to be satisfied by all materials presently known to occur in nature, in my opinion it is not at all a fundamental aspect of thermodynamics. Indeed, there are entirely sensible models of quantum dynamical systems (such as a quantum Boltzmann gas) which violate it (see, e.g., Huang (1963) for further discussion). Thus, I do not view the failure of the analog of this law to hold in black hole physics as particularly disturbing.

There is an alternate—and much vaguer—formulation of the third law, which states that it is impossible to achieve absolute zero temperature in a finite series of processes. (I would view this statement as being, in essence, a consequence of the second law, since it is entropically highly unfavorable to remove energy from a system near absolute zero temperature.) It appears that an analog of this formulation of the third law does hold in black hole physics (see Israel (1986)).

Taken together, the above results provide a striking mathematical analogy between the laws of thermodynamics and classical black hole physics. Furthermore, there even is a hint that the analogy may have some physical content as well: The quantity in the laws of black hole physics which plays the role mathematically analogous to total energy, E, is the mass, M, of the black hole, which, in general relativity, physically *is* the total energy of the black hole. However, at this stage, the physical analogy ends. The quantity in black hole physics mathematically analogous to temperature, T, is the surface gravity, κ. In classical black hole physics, κ has nothing whatever to do with the physical temperature of a black hole, which is absolute zero by any reasonable criteria (such as the ability to run heat engines). Nevertheless, we shall see in the next chapter, that this situation changes dramatically when quantum physics is taken into account.

7 The Hawking Effect

In chapter 5, we saw that in extended Schwarzschild spacetime, the Hartle-Hawking vacuum state—which is uniquely characterized by being everywhere nonsingular (i.e., of Hadamard form) and invariant under the Schwarzschild time translation isometries—is a thermal state at the "Hawking temperature" $\kappa/2\pi$. However, we also noted there that the Hartle-Hawking state has the rather unphysical feature of having incoming thermal radiation in region I of extended Schwarzschild spacetime correlated with similar incoming radiation in region II, as well as with radiation emerging from the "white hole" singularity. It is far from clear, *a priori*, what the relationship is between the mathematical properties of the Hartle-Hawking state in extended Schwarzschild spacetime and the physical effects that would occur when a black hole is produced in nature by the gravitational collapse of a body. Nevertheless, we shall see in section 7.1 that the analysis of the behavior of a quantum field at sufficiently late times following gravitational collapse to a black hole is closely related mathematically to the analysis of the Unruh effect given in chapter 5. The result of the analysis is the *Hawking effect* : A black hole will radiate exactly like a blackbody at temperature $\kappa/2\pi$.

One immediate ramification of the Hawking effect is that it solidifies the relationship between black holes and thermodynamics presented in the previous chapter. The surface gravity, κ, is not merely a mathematical analog of temperature, it literally is the physical temperature of a black hole. This strongly suggests that one-fourth of the area of a black hole should be interpreted as its physical entropy. These ideas are discussed in section 7.2.

A further major ramification of the Hawking effect arises when one considers the back-reaction of the quantum field on the black hole. As a result of back-reaction effects, an isolated black hole should completely "evaporate" within a finite time. Although a complete analysis of this phenomenon will have to await a quantum theory of gravity, it would seem difficult to avoid the conclusion that, in this process, an initially pure quantum state will evolve to a mixed state, i.e., "loss of quantum coherence" should occur. This ramification of the Hawking effect is explored in section 7.3.

7.1 Particle Creation by Black Holes

We wish to consider quantum field effects in a spacetime corresponding to the gravitational collapse of a body to form a black hole. For simplicity, we shall restrict attention to the case of spherically symmetric collapse to a Schwarzschild black hole. As we shall argue later, the behavior of the quantum field at asymptotically late times depends only upon the asymptotic final state of the black hole, so a nonspherical collapse producing a black hole which merely "settles down" to a Schwarzschild black hole at late times would yield the same results. The modifications to the analysis occurring for the case of collapse to a Kerr black hole (where the phenomenon of "superradiance" occurs) will not be discussed here, but can be found in Hawking (1975) and Wald (1975).

The spacetime geometry corresponding to the gravitational collapse of an initially static, spherical body to a Schwarzschild black hole is shown in fig. 7.1. Since the spacetime geometry is (by assumption) spherically symmetric, by Birkhoff's theorem, the region exterior to the body is isometric to a portion of extended Schwarzschild spacetime. Note, however, that this exterior region corresponds to only a part of regions I and III of fig. 5.3, and analogs of regions II and IV of that figure are entirely absent from fig. 7.1. In particular, no analog of the past horizon, $\mathfrak{h}_B$, appears in fig. 7.1.

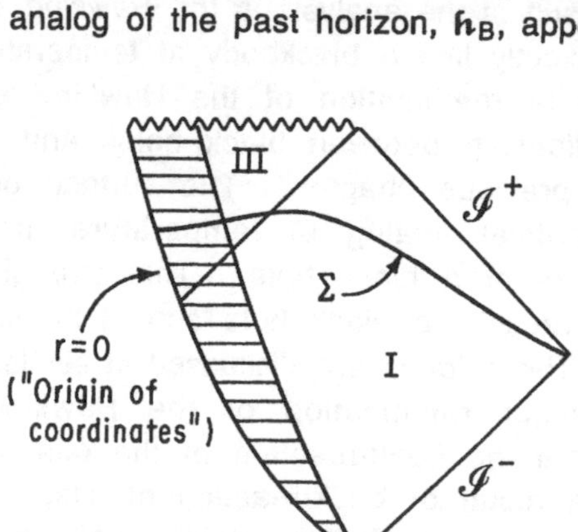

Figure 7.1. A conformal diagram of a spacetime in which a spherical body undergoes collapse to a Schwarzschild black hole.

Consider, now, a Klein-Gordon quantum field in the spacetime of fig. 7.1. Since this spacetime is asymptotically static in the past, we have a natural quantum field theory construction—the "in" representation—which is obtained in the manner described at the end of section 4.3. States in the Fock space of the "in" representation have a natural particle interpretation in the asymptotic past.

Although the spacetime of fig. 7.1 also is static in region I in the asymptotic future, the static portion of region I does not contain a Cauchy surface for the spacetime, since any Cauchy surface must pass through the collapsing matter. Consequently, the construction of the "out" representation is less straightforward, and we proceed as follows: First, we assume that in the asymptotic future, each $\psi \in \mathcal{S}$ propagates entirely through the black hole horizon, $\mathfrak{h}_A$, and/or out to infinity. This "asymptotic completeness" assumption is supported by analyses of scattering in Schwarzschild spacetime (Dimock and Kay (1987); see also Wald (1984a)) but, as far as I am aware, has been rigorously proven only in the massless case (Dimock 1985). We then focus attention upon a subspace, $\mathcal{S}_L$, of $\mathcal{S}$, comprised by solutions which propagate into the black hole and/or to infinity at sufficiently "late times". For the spacetime of fig. 7.1, we make the following choice of $\mathcal{S}_L$: Let Σ be any Cauchy surface which intersects the event horizon outside of the collapsing matter and let $\mathcal{S}_L$ consist of the solutions in $\mathcal{S}$ whose data on Σ have support outside the black hole. (In the more general case where the spacetime geometry merely "settles down" to a Schwarzschild final state, Σ should be chosen so that the spacetime geometry has approached its Schwarzschild final state throughout the future of Σ in region I, so that the spacetime can be considered to be static when solutions in $\mathcal{S}_L$ propagate into the black hole and/or to infinity.) We define the "early time" subspace, $\mathcal{S}_E$, of $\mathcal{S}$, to be the symplectic complement of $\mathcal{S}_L$ in $\mathcal{S}$, so that $\mathcal{S}_E$ consists of solutions in $\mathcal{S}$ whose data on Σ have support inside the black hole. As we shall see below, the solutions in $\mathcal{S}_E$ will play a role in our analysis similar to the role played by solutions in region II of Minkowski spacetime in the derivation of the Unruh effect.

Since the portion of the spacetime of fig. 7.1 outside the collapsing matter is isometric to a portion of extended Schwarzschild spacetime, we may identify each solution in $\mathcal{S}_L$ with the solution in extended Schwarzschild spacetime which has the same initial data on a Cauchy surface $\tilde{\Sigma}$ of extended Schwarzschild spacetime which coincides with Σ in region I under the identification provided by the isometry relating the two spacetime regions. Given $\psi_L \in \mathcal{S}_L$, we denote the corresponding solution in extended Schwarzschild spacetime by $\tilde{\psi}_L$. We define the "out" representation to be such that for all $\psi_L \in \mathcal{S}_L$, the associated "mode function", $K\psi_L$, in the one-particle Hilbert space, $\mathcal{H}_{out}$, is the solution in the spacetime of fig. 7.1 which corresponds to the positive frequency part (with respect to Schwarzschild time translations) of $\tilde{\psi}_L$ in extended Schwarzschild spacetime. (Note that the positive frequency part of $\tilde{\psi}_L$ in extended Schwarzschild spacetime will have a small "tail" which propagates into the black hole at "early times," resulting in some ambiguity in identifying it with a solution in the spacetime of fig. 7.1. However, for solutions $\psi_L \in \mathcal{S}_L$ which propagate into the black hole or out to infinity at sufficiently late times, this "tail"—and hence this ambiguity in identification—becomes arbitrarily small.) We denote by $\mathcal{H}_L$ the (complex) subspace, $K\mathcal{S}_L$, of $\mathcal{H}_{out}$ spanned by the mode functions associated with $\mathcal{S}_L$. We extend K to $\mathcal{S}$ and we write $\mathcal{H}_E = K\mathcal{S}_E$. We shall leave the choice of extension of K to $\mathcal{S}$ unspecified for the present, but in the course of our analysis below we will make a convenient choice of $K\psi_E$ for certain $\psi_E \in \mathcal{S}_E$. It should be emphasized that the "out" representation Fock space we have just defined has a physically meaningful particle interpretation for modes in $\mathcal{H}_L$ (with respect to measurements by static observers in the future), but no meaningful particle interpretation is claimed for modes in $\mathcal{H}_E$.

We wish, now, to compute the S-matrix, $U: \mathcal{F}_s(\mathcal{H}_{in}) \to \mathcal{F}_s(\mathcal{H}_{out})$ relating the "in" and "out" representations. In particular, we wish to calculate $U|0_{in}\rangle$. The density matrix in $\mathcal{F}_s(\mathcal{H}_L)$ corresponding to $U|0_{in}\rangle$ will then tell us about the spontaneous creation of particles as seen by static observers at late times. We shall give a sketch of the calculation of $U|0_{in}\rangle$ along the lines of the original derivation (Hawking 1975, Wald, 1975) and then will describe an improved framework for analyzing this effect.

The relevant information needed to compute the density matrix in $\mathcal{F}_s(\mathcal{H}_L)$ corresponding to $U|0_{in}\rangle$ can be obtained by starting with solutions in $\mathcal{H}_L$ (which, as described above, are "positive frequency" in region I in the future), propagating them into the past, and decomposing them into their positive and negative frequency parts there. We now assume the same "asymptotic completeness" property in extended Schwarzschild spacetime for the propagation of classical solutions into the past as we assumed above for the propagation of classical solutions into the future in the spacetime of fig. 7.1, i.e., we assume that in the asymptotic past, all solutions in extended Schwarzschild spacetime propagate entirely into the white hole and/or out to infinity. We then may further decompose $\mathcal{H}_L \subset \mathcal{H}_{out}$ as

$$\mathcal{H}_L = \mathcal{H}_{wh} \oplus \mathcal{H}_{-\infty} \tag{7.1.1}$$

where $\mathcal{H}_{wh}$ consists of solutions, ψ_L, which are positive frequency in the future and whose corresponding solution $\tilde{\psi}_L$ in extended Schwarzschild spacetime propagates entirely through the "white hole" horizon in the past. Similarly, $\mathcal{H}_{-\infty}$ consists of solutions, $\psi_L \in \mathcal{H}_L$, such that the corresponding solution $\tilde{\psi}_L$ propagates entirely to infinity in the asymptotic past. Now, apart from a "tail" which becomes arbitrarily small at late times, the solutions in $\mathcal{H}_{-\infty}$ never propagate into the nonstatic region of the spacetime of fig. 7.1, i.e., their propagation is exactly the same as occurs in extended Schwarzschild spacetime. It follows immediately that for all $\psi_{-\infty} \in \mathcal{H}_{-\infty}$, there is no "mixing of frequencies", so we have

$$DC^{-1}\psi_{-\infty} = 0 \tag{7.1.2}$$

This implies that no particle creation occurs in modes corresponding to $\mathcal{H}_{-\infty}$. Note that this contrasts sharply with the properties of the Hartle-Hawking vacuum state of extended Schwarzschild spacetime, in which all modes in region I (including those corresponding to $\mathcal{H}_{-\infty}$) are thermally populated.

Consider, now, the propagation of a solution $\psi_{wh} \in \mathcal{H}_{wh}$ in the spacetime of fig. 7.1 such that ψ_{wh} emerges to infinity and/or enters the black hole at sufficiently late times in the future. (Note that since $\psi_{wh} \in \mathcal{H}_{wh} \subset \mathcal{H}_L$, ψ_{wh} automatically emerges to infinity and/or enters the black hole at "late times", i.e., within the static Schwarzschild region of the spacetime of fig. 7.1. By "sufficiently late times", we mean at sufficiently large values of Killing parameter within this static region.) When evolved backwards into the past, ψ_{wh} will propagate through the collapsing matter and emerge in the asymptotic past as an ultra-high-frequency solution. (One way of seeing this is that if one considers the "time translates" of ψ_{wh} to increasingly later times, the "blueshift" of these solutions as determined by any fixed, smooth family of observers falling into the black hole continues to increase, but the subsequent "redshift" resulting from their continued propagation through the collapsing matter approaches a finite limit.) We wish to decompose ψ_{wh} into its positive and negative frequency parts in the asymptotic past. The key result needed to do so is the following: The positive and negative frequency decomposition of ψ_{wh} in the asymptotic past in the spacetime of fig. 7.1 is the same as the decomposition of the corresponding solution, $\tilde{\psi}_{wh}$, in extended Schwarzschild spacetime, where "positive and negative frequencies" are defined with respect to *affine parameter* along the white hole horizon, $\mathfrak{h}_B$. This follows from the fact that the geometrical optics approximation can be used to determine the propagation of ψ_{wh} into the asymptotic past starting from the Schwarzschild region just outside the collapsing matter, since ψ_{wh} is "ultra-high-frequency" throughout this region. Consequently, in this region, the surfaces of constant phase of the wave will be null. The propagation of the "pattern" of the wave will be governed by the geodesic deviation equation applied to the congruence of null geodesics which generate the surfaces of constant phase, and the above result then follows (see Hawking (1975) or Wald (1984a) for further details). Now, $\tilde{\psi}_{wh}$ is purely positive frequency with respect to Schwarzschild time translations, so, to complete the calculation, we simply need the relationship between the positive and negative frequency decompositions of functions on the white hole horizon, $\mathfrak{h}_B$, of extended Schwarzschild spacetime, with respect to "Killing time" and "affine time". This, of course, was the key ingredient which entered our analysis of the Unruh effect. From eqs. (5.1.20) and (5.1.21) and their generalization described in section 5.3, we see that if $\tilde{\psi}_{wh}$ is peaked sharply around the frequency $\omega > 0$, then in extended Schwarzschild spacetime the solutions

$$\tilde{\Psi}_\omega = \tilde{\psi}_{wh} + \exp(-\pi\omega/\kappa)\, \overline{\tilde{\psi}}_{II} \qquad (7.1.3)$$

and

$$\tilde{\Psi}'_\omega = \tilde{\psi}_{II} + \exp(-\pi\omega/\kappa)\, \overline{\tilde{\psi}}_{wh} \qquad (7.1.4)$$

are purely positive frequency with respect to affine parameter on the white hole horizon, $\mathfrak{h}_B$, where $\tilde{\psi}_{II}$ is the solution (whose initial data has support in region II of extended Schwarzschild spacetime) obtained by applying the "wedge reflection" isometry to $\tilde{\psi}_{wh}$, followed by complex conjugation. Consequently, in the spacetime of fig. 7.1, the solutions

$$\Psi_\omega = \psi_{wh} + \exp(-\pi\omega/\kappa)\ \overline{\psi}_{II} \tag{7.1.5}$$

$$\Psi'_\omega = \psi_{II} + \exp(-\pi\omega/\kappa)\ \overline{\psi}_{wh} \tag{7.1.6}$$

are purely positive frequency in the asymptotic past, where ψ_{II} is the solution corresponding to $\tilde{\psi}_{II}$ in the "asymptotic future" (inside the black hole). Note that in the spacetime of fig. 7.1, ψ_{II} enters the black hole at "early times".

It is convenient, now, to define the "early time mode" subspace, $\mathcal{H}_E$, of the "out" one-particle Hilbert space, $\mathcal{H}_{out}$, to be such that for all ψ_{wh} which emerge to infinity and/or enter the black hole at asymptotically late times, we have $\psi_{II} \in \mathcal{H}_E$. (The final results for the density matrix on $\mathcal{F}_s(\mathcal{H}_L)$ do not depend on this choice.) Then from eqs. (7.1.5) and (7.1.6) we find that—exactly as in the derivation of eqs. (5.1.24) and (5.1.25) above—for any $\psi_{wh} \in \mathcal{H}_L$ which is peaked sharply around frequency $\omega > 0$, we have

$$DC^{-1}\psi_{wh} = \exp(-\pi\omega/\kappa)\ \overline{\psi}_{II} \tag{7.1.7}$$

$$DC^{-1}\psi_{II} = \exp(-\pi\omega/\kappa)\ \overline{\psi}_{wh} \tag{7.1.8}$$

Consequently, by an exact repetition of the steps which led to eq. (5.1.28), we obtain the following remarkable conclusion, known as the *Hawking effect*: *The state $U|0_{in}>$ is a thermal state at the Hawking temperature $T = \kappa/2\pi$ with respect to modes in $\mathcal{H}_{wh}$ which emerge to infinity at asymptotically late times. It is a vacuum state with respect to modes in $\mathcal{H}_{-\infty}$.*

These conclusions can be formulated more succinctly by stating simply that, at late times, a Schwarzschild black hole formed by gravitational collapse radiates precisely as a thermal blackbody at temperature

$$kT = \kappa/2\pi = \frac{\hbar c^3}{8\pi G M} \tag{7.1.9}$$

Namely, the modes in $\mathcal{H}_{-\infty}$ correspond to particles which appear to have propagated in from infinity at late times; these particle states are unpopulated. On the other hand, the modes in $\mathcal{H}_{wh}$ correspond to particles emanating from the white hole in the extended Schwarzschild spacetime. In the spacetime of fig. 7.1, these particles would appear to be emerging from the region where the black hole has formed. These particle states are thermally populated at the temperature (7.1.9). Thus, the net effect is radiation identical in nature to that which would be emitted by a perfect blackbody radiating into empty space (see section 7.3 of Dimock and Kay (1987)).

In extended Schwarzschild spacetime, one also could consider the state corresponding to the late time behavior of the quantum field in the spacetime of fig. 7.1. In region I, this state—known as the *Unruh vacuum*—would be a vacuum state with respect to particles propagating in from past infinity and a thermal state at the Hawking temperature with respect to particles emanating from the white hole. However, the Unruh vacuum would not be a physically acceptable state in extended Schwarzschild spacetime, since, as already mentioned in section 5.3, it is singular on the white hole horizon $\mathfrak{h}_B$. (That the Unruh vacuum must be singular somewhere follows from the general uniqueness result described in section 5.3, which shows that the Hartle-Hawking vacuum is the only globally nonsingular state which is invariant under the Schwarzschild time translations.) There is no analogous difficultly in the spacetime of fig. 7.1, since no analog of $\mathfrak{h}_B$ exists there.

It is worth noting that although the above prediction that a black hole radiates as a blackbody holds only at "sufficienly late times", an examination of the derivation shows that observers near infinity should see this thermal radiation for all times, t, such that $t - t_0 >> t_D$, where t_0 denotes the "time of black hole formation" (defined in Wald (1976)) and the "dynamical timescale", t_D, is given by

$$t_D \sim \frac{GM}{c^3} \sim 10^{-5}\frac{M}{M_\odot} \text{ sec} \tag{7.1.10}$$

Thus, in fact, a black hole approaches its perfect blackbody state very rapidly.

Although a few of the steps in the construction of $\mathcal{H}_{\text{out}}$ and other aspects of the above derivation of the Hawking effect have been argued somewhat loosely, I see no obstacle to making this derivation completely rigorous, provided only that suitable forms of asymptotic completeness can be proven for classical scattering both in the spacetime of fig. 7.1 and in extended Schwarzschild spacetime. Nevertheless, the derivation we have presented is quite awkward in the following two respects. First, as in the derivation of the Unruh effect, the S-matrix, U, does not actually exist, i.e., the "in" and "out" representations defined above are unitarily inequivalent. (Since the formal expression for U can still be interpreted in the manner described near the end of section 4.5, this does, indeed, correspond to an "awkwardness" of the derivation rather than a "fatal flaw".) Second, the results of the analysis concern the behavior of the quantum field at sufficiently late times, but the notion of particle is inherently a nonlocal one, i.e., the particle modes which propagate out to infinity and/or enter the black hole at "sufficiently late times" have small "tails" which reach infinity or enter the black hole at "early times". Consequently, one cannot give a clean, simple definition of $\mathcal{H}_L$. As already noted above, this is not a serious problem because these "tails" become arbitrarily small for modes which propagate to infinity and/or enter the black hole at sufficiently late times. Nevertheless, it is awkward to describe the late time behavior of the field in terms of such particle modes.

A much more sensible approach to the analysis of particle creation by black holes—which avoids these awkward features of the S-matrix derivation—would be to describe the phenomenon entirely in terms of the local behavior of the correlation functions, $<\hat{\phi}(x_1) \cdots \hat{\phi}(x_n)>$, of the quantum field. In other words, we may view the problem as an evolution problem for these correlation functions: Starting with the correlation functions for the quantum field in the static vacuum state (or some other physically reasonable state) in the past, what is the behavior of the correlation functions at late times in the future? Such an approach has been taken by Fredenhagen and Haag (1990) and we now briefly describe the main features of their analysis.

Let Σ be a Cauchy surface for the spacetime of fig. 7.1 which intersects the black hole event horizon outside of the collapsing matter. (As indicated previously, if the collapse were nonspherical, then Σ should be chosen so that it intersects the event horizon after the black hole has "settled down" to its stationary final state.) Now, for any test functions $f_1,\ldots,f_n \in \mathcal{T}$, and any state, ω, we have (see eq. (3.2.28))

$$<\hat{\phi}(f_1) \cdots \hat{\phi}(f_n)>_\omega = <\hat{\Omega}(\psi_1,\cdot) \cdots \hat{\Omega}(\psi_n,\cdot)>_\omega \qquad (7.1.11)$$

where $\psi_i = Ef_i$, with E being the advanced minus retarded Green's function, eq. (3.2.22). By the construction used in the proof of the first part of lemma 3.2.1, the right side of eq. (7.1.11) may then be re-expressed in terms of the n-point correlation function smeared with test functions with support in an arbitrarily small neighborhood of Σ. In this manner, the correlation functions of the field can be obtained by "evolution" from their values in a neighborhood of Σ.

We wish to determine the correlation functions of the field in region I of fig. 7.1 at asymptotically late times. To do so, we start with the correlation function (7.1.11) for any test functions f_i having support in the exterior region to the future of Σ, and we consider the behavior of this correlation function as these test functions are "time translated" via the Schwarzschild time translations by time t, with t large and positive. The corresponding solutions $\psi_i = Ef_i$ also will be "time translated" by t. The basic behavior of the initial data for these time translated solutions on Σ in the limit as $t \to \infty$ then can be seen from the following argument: If we assume, again, that a suitable form of asymptotic completeness holds in the extended Schwarzschild spacetime of fig. 5.3, then the corresponding solution $\tilde{\psi}_i$ in extended Schwarzschild spacetime will be characterized by a portion which propagates in from the "white hole" and a portion which propagates in from infinity. When "time translated" by t, data for $\tilde{\psi}_i$ on the white hole horizon, $\mathfrak{h}_B$, will be translated "upwards" toward the bifurcation surface S (since the flow of the Schwarzschild isometries on $\mathfrak{h}_B$ is toward S); similarly, the portion of $\tilde{\psi}_i$ propagating in from infinity will emerge from infinity at a later time. This implies that the initial data on a corresponding Cauchy surface, $\tilde{\Sigma}$, in extended Schwarzschild spacetime for the time translate of $\tilde{\psi}_i$ by t will have support which is concentrated, in the limit as $t \to \infty$, arbitrarily near spatial infinity and arbitrarily near the black hole horizon. Since the exterior portion of the spacetime of fig. 7.1 lying to the future of Σ is isometric to the corresponding portion of extended Schwarzschild spacetime, the analogous conclusion obviously holds for the spacetime of fig. 7.1. This implies that in a spacetime describing gravitational collapse to a black hole, the correlation functions of the quantum field at asymptotically late times are determined by the correlation functions at "time" Σ in a neighborhood of spatial infinity and in an arbitrarily small neighborhood of the event horizon of the black hole.

To proceed, Fredenhagen and Haag (1990) assume that the two-point function of the quantum field in the vicinity of Σ near spatial infinity agrees with that of the Boulware vacuum, i.e., the static vacuum state in region I of extended Schwarzschild spacetime. This corresponds to assuming that no incoming particles are sent in from infinity at sufficiently late times. It follows immediately that the two-point function continues to agree with that of the Boulware vacuum at late times when smeared with test functions corresponding to solutions which do not propagate through the collapsing matter. (This corresponds to the conclusion of the analysis given above that no particle creation occurs for modes in $\mathcal{H}_{-\infty}$.) They then further assume that near the black hole horizon, the two-point function has leading order singularity structure which agrees with the Hadamard form. This assumption is well motivated, since as discussed in section 4.6, the Hadamard form of the two-point function is necessary for the quantum field to have a nonsingular stress-energy tensor. Furthermore, since the vacuum state in a static spacetime has Hadamard form (Fulling et al. 1981) and since Hadamard form is preserved under evolution (Fulling et al. 1978; Kay and Wald 1991), the Hadamard form of the "in" vacuum state automatically holds throughout the spacetime. Thus, the leading order behavior of the two-point function on Σ near the horizon is the same as that of the Hartle-Hawking vacuum of extended Schwarzschild spacetime, which is a thermal state at the temperature (7.1.9) with respect to the Schwarzschild time translations (see section 5.3). Fredenhagen and Haag then show that, when smeared at sufficiently late times with test functions corresponding to solutions which propagate through the collapsing matter (i.e., solutions, ψ_L, such that $K\psi_L$ lies in $\mathcal{H}_{wh}$), the two-point function corresponds to a thermal state at temperature (7.1.9). Presumably, an extension of their analysis could be used to show that at sufficiently late times throughout region I, all the correlation functions of the field correspond to having a thermal population of the modes in $\mathcal{H}_{wh}$ and a vacuum state for all the modes in $\mathcal{H}_{-\infty}$. In this manner, the Hawking effect can be derived in a manner that avoids the awkward features of the S-matrix derivation.

The derivation of Fredenhagen and Haag sketched above also clarifies several important features of the particle creation calculation. First, the derivation makes it manifest that the conclusion of thermal emission at late times depends only upon the assumption that there is no incoming radiation from infinity at late times and that the two-point function of the state has leading order singularity structure of Hadamard form near the horizon at "time" Σ. (In order to get thermal behavior of all of the correlation functions at late times, it presumably also would be necessary to assume additional nonsingularity conditions on them at "time" Σ.) Since the Hadamard form is preserved under evolution, this latter assumption holds if the state is nonsingular at arbitrarily early times. Thus, the thermal emission property actually holds for *all* physically nonsingular incoming states, not merely for $|0_{in}>$. (A similar—but more awkwardly formulated—conclusion was obtained in Wald (1976) from S-matrix arguments.) Secondly, since the derivation involves only evolution properties to the future of Σ, it is made manifest that the late time emission depends only on the final state of the black hole; the detailed nature of the collapse and the manner in which the black hole "settles down" to its final state are not relevant.

Finally, we note that the derivation of Fredenhagen and Haag also brings to the fore one disturbing feature of the derivation of particle creation by black holes: The derivation of thermal behavior of the quantum field at asymptotically late times is seen to arise from the singularity structure of the two-point function at arbitrarily short distances. However, even ignoring possible new effects arising from the quantum nature of gravity itself at distance scales smaller than the Planck length, it is unreasonable to assume that the simple, linear field model considered in the derivation will provide an accurate model to a realistic field theory at ultra-short-distance scales. Thus, one might question whether the particle creation effect will occur for nonlinear (i.e., interacting) fields even if these fields can be treated as noninteracting on large distance scales (i.e., at "low energies").

In response to this issue, it should be noted that the derivation of the Unruh effect for linear fields discussed in section 5.1 similarly depends upon the ultra-short-distance singularity structure of the two-point function near the horizon. Nevertheless, the theorem of Bisognano and Wichmann (1976) shows that in Minkowski spacetime, the Unruh effect continues to hold for nonlinear fields. Furthermore, the arguments given at the end of section 5.3 strongly suggest that the Unruh effect also continues to hold for nonlinear fields in static curved spacetimes. For this reason—and also because of the intimate relationship between black holes and thermodynamics discussed further in the next section—I see no reason to doubt that, even for interacting fields, a black hole will continue to emit in exactly the same manner as a perfect blackbody; see

Jacobson (1993) for further recent discussion of this issue. Note that since nonlinear interactions will couple the (initially thermally populated) modes in $\mathcal{H}_{wh}$ to the (initially unpopulated) modes in $\mathcal{H}_{-\infty}$, the actual spectrum of particles seen by a distant observer should deviate from a thermal one in a manner that would depend upon the details of the nonlinear interactions. This same remark would apply, of course, to an ordinary blackbody which radiates into empty space.

7.2 The Generalized Second Law

One of the most remarkable implications of the Hawking effect is the cementing of the relationship between the laws of black hole mechanics and the laws of thermodynamics. We saw in section 6.2 that at the classical level, there is a striking mathematical analogy between the ordinary laws of thermodynamics and certain laws applying to black holes. As we noted at the end of section 6.2, in this correspondence of laws, the mass of the black hole plays the same mathematical role as the total energy of a thermodynamic system. Since mass and energy represent the same physical quantity, this suggests that the analogy of laws might have some physical content. However, classically this physical analogy breaks down: The quantity in black hole physics which plays the role mathematically analogous to temperature in thermodynamics is the surface gravity, κ, but the physical temperature of a classical black hole is absolute zero. However, the results of the previous section produce a dramatic change in this situation: *The Hawking effect shows that $\kappa/2\pi$ truly is the physical temperature of a black hole.* Hence, this suggests the possibility that the laws of black hole mechanics truly are the ordinary laws of thermodynamics applied to a system containing a black hole.

If we compare eqs (6.2.15) and (6.2.23) making the physical identifications $M \leftrightarrow E$ and $\kappa/2\pi \leftrightarrow T$, we see that the mathematical analogy between the laws of black hole physics and the laws of thermodynamics would carry over to a complete physical analogy if we could physically identify $A/4$ and S, i.e., if $A/4$ were to represent the physical entropy of a black hole. Thus, the key remaining issue in black hole thermodynamics is whether the physical entropy of a black hole is given by

$$S_{bh} = \frac{kc^3}{4G\hbar} A \qquad (7.2.1)$$

where, in this formula, we have restored the fundamental constants of nature G, c, and $\hbar$ as well as Boltzmann's constant, k. Note that eq. (7.2.1) corresponds to an enormous entropy. In particular, for a black hole of a solar mass, eq. (7.2.1) yields $S_{bh}/k \sim 10^{77}$. By comparison, the ordinary entropy of the Sun is very roughly $S_{sun}/k \sim 10^{58}$.

One approach to determine whether eq. (7.2.1) is valid would be to attempt to directly calculate the entropy of a black hole from first principles. However, this would require a detailed description of the microscopic degrees of freedom of a black hole at the quantum level, i.e., it should require us to have a complete, quantum theory of gravity. We do not have such a theory at present. Nevertheless, an interesting derivation of eq. (7.2.1) from ideas suggested by a Euclidean approach to quantum gravity was given by Gibbons and Hawking (1977b). For an ordinary quantum system, the partition function, Z, is defined by

$$Z = \operatorname{tr} e^{-\beta H} \tag{7.2.2}$$

where H is the Hamiltonian operator. An argument using the canonical (or grand canonical ensemble) then establishes that

$$S = \ln Z + \beta E \tag{7.2.3}$$

(Additional terms involving chemical potentials will be present on the right side if there are other conserved quantities besides energy for the system plus heat bath.) For an ordinary quantum system, the right side of eq. (7.2.2) may be calculated using path integral methods. One obtains

$$\operatorname{tr} e^{-\beta H} = \int \mathcal{D}[\text{paths}]\, e^{-\mathcal{S}_E} \tag{7.2.4}$$

where $\mathcal{S}_E$ denotes the "Euclidean action", and the integral is taken over all Euclidean paths which are periodic in Euclidean time with period β.

In quantum field theory, one commonly evaluates (approximately) the right side of eq. (7.2.4) by expanding about a minimum of $\mathcal{S}_E$ and then calculating the contribution to Z in the "one-loop approximation". In the case of the black hole, Gibbons and Hawking (1977b) calculated the path integral (7.2.4) in the "zero-loop approximation" by simply evaluating $\mathcal{S}_E$ on a "quasi-Euclidean section." Remarkably, they then found that the entropy derived from the partition function in this approximation is given precisely by eq. (7.2.1). More generally (i.e., when other matter fields are included or in other theories of gravity), it can be shown that this method of calculation gives results in agreement with the method outlined in section 6.2 (Wald 1993c).

The result of this calculation is supportive both of eq. (7.2.1) and of the Euclidean approach to quantum gravity. However, there is at least one disturbing aspect of the calculation. From eq. (7.1.9) it is readily seen that the temperature of a Schwarzschild black hole varies inversely with its mass-energy, i.e., a Schwarzschild black hole has a negative heat capacity. This result should not be surprising on physical grounds, since an ordinary self-gravitating star in Newtonian gravity also has a negative heat capacity; if one removes energy from a star, it contracts and heats up. As in the case of an ordinary star, this negative heat capacity does not imply any fundamental difficulty in describing the thermodynamics of black holes, since the microcanonical ensemble still should be well defined for a finite system containing a black hole, and a black hole can exist in stable, thermal equilibrium in a sufficiently small box with walls that perfectly reflect radiation. However, the negative heat capacity implies that a Schwarzschild black hole cannot exist in a stable thermal equilibrium with an ordinary heat bath (at fixed "temperature as measured at infinity"). But such an equilibrium should be necessary in order to justify the use of the canonical ensemble for a black hole in the calculation given above. This problem also manifests itself by the fact that $A = 16\pi M^2$ for a Schwarzschild black hole, so—assuming the usual interpretation of entropy—the density of states of a Schwarzschild black hole should grow with M as $\exp(4\pi M^2)$. However, in that case, the sum, eq. (7.2.2), defining Z would not converge. Thus, there appears to be a logical inconsistency in the above Euclidean path integral calculation of S_{bh}, since the result of the calculation would seem to invalidate the method used to derive it. Some ideas as to how to overcome these difficulties by redefining the canonical ensemble have been suggested by York and collaborators (see Braden et al. (1987), Brown et al. (1990), and references cited therein). More recently, Brown and York (1993a,b) have proposed a derivation of these results based upon the microcanonical ensemble. Nevertheless, the essentially classical (i.e., "zero-loop") nature of the Euclidean path integral derivation of the formula $S_{bh} = A/4$ remains rather mysterious.

Although it is not likely that one will be able to improve upon the above calculation of S_{bh} from first principles until a complete quantum theory of gravity is at hand, there is a further idea—known as the "generalized second law"—which strongly suggests that A/4 must be regarded as the physical entropy of a black hole. To explain this idea, it first should be noted that there are some difficulties both with the ordinary second law of thermodynamics and with the area theorem. A difficulty with the ordinary second law arises when a black hole is present: One can take some matter and dump it into a black hole in which case—at least, according to classical general relativity—it will disappear into the singularity within the black hole. In this manner, the total entropy of matter in the universe can be decreased. On the other hand, the area theorem clearly must be violated in the quantum particle creation process, since the mass, M, of a Schwarzschild black hole (and hence its area $A = 16\pi M^2$) must decrease in the process if energy is to be conserved. (This violation of the conclusion of theorem 6.1.2 can occur because the expected stress-energy tensor of the quantum field violates the null energy condition at the horizon of the black hole.) Indeed, as we shall discuss in the next section, "back-reaction" effects of particle creation are expected to reduce the area of an isolated black hole to zero within a finite time. Note, however, that when the total entropy, S_m, of matter outside of black holes is decreased by dumping matter into a black hole, A will tend to increase. Similarly, when A is decreased during the particle creation process, thermal matter is created outside the black hole, so S_m increases. Thus, although S_m and A each can decrease individually, it is possible that the *generalized entropy*, S', defined by

$$S' = S_m + \frac{1}{4}A \qquad (7.2.5)$$

never decreases. The conjecture that $\Delta S' \geq 0$ in all processes was first put forth by Bekenstein (1974)—prior to the discovery of the Hawking effect!—and is known as the *generalized second law*.

If valid, the generalized second law would have a very natural interpretation: Presumably, it simply would be the ordinary second law applied to a system containing a black hole. If so, then there could be no question that A/4 truly represents the physical entropy of a black hole. Thus, a key issue in the subject of black hole thermodynamics is whether the generalized second law holds.

At first thought, one might expect that the generalized second law could be violated easily as follows. For simplicity, consider the case of a static black hole. In that case, the Killing field which is timelike at infinity coincides with the Killing field, χ^a, which is normal to the horizon of the black hole. Far from the black hole, put matter of energy E and entropy S into a box and then lower the box on a rope toward the black hole. When the horizon is reached, open the box and allow the matter to fall into the black hole. Since no entropy need be generated in the lowering process, the entropy of matter outside the black hole will be decreased by S in this process, i.e., $\Delta S_m = -S$. On the other hand, the area change of the black hole can be calculated as follows. The force exerted by a distant observer who holds the rope is given by

$$F_\infty = E \frac{d\chi}{dy} \tag{7.2.6}$$

where χ denotes the "redshift factor" at the box (i.e., $\chi^2 = -\chi_a \chi^a$) and y denotes proper distance along the path (orthogonal to the surfaces of constant χ) followed by the box in the static hypersurface. (Here, for simplicity, it is assumed that the height (i.e., y-extent) of the box is negligible.) Hence, the work done at infinity during the process of lowering the box is

$$W_\infty = \int F_\infty \, dy = (1 - \chi) E \tag{7.2.7}$$

Thus, by conservation of energy, the energy delivered to the black hole is

$$\Delta M = E - W_\infty = \chi E \tag{7.2.8}$$

By the first law of black hole mechanics, (see eq. (6.2.8) or eq. (6.2.15)) the area increase of the black hole in this process is given by

$$\Delta A = \frac{8\pi}{\kappa} \Delta M = \frac{8\pi}{\kappa} \chi E \tag{7.2.9}$$

However, at the horizon we have $\chi = 0$, so by lowering the box sufficiently close to the horizon, we can make ΔA arbitrarily small. Thus, it would appear that we can make $\Delta S' = -S + \Delta A/4$ negative, in violation of the generalized second law.

The above calculation does not take quantum effects like Hawking radiation into account, but since the calculation is applicable to a macroscopically large black hole, one might expect that such quantum effects could be made negligible simply by taking M to be sufficiently large. Remarkably, however, although for a large black hole the Hawking radiation seen at infinity indeed is negligible and there are no important nonclassical effects on freely falling bodies, we now shall show that the Unruh effect makes a large quantum correction to the behavior of a body which is slowly lowered toward the horizon of the black hole. As discussed in section 5.3, when the field is in the natural vacuum state of the extended black hole spacetime, a static observer will see himself immersed in a thermal bath at the locally measured temperature,

$$T = \frac{\kappa}{2\pi\chi} \tag{7.2.10}$$

(see eq. (5.3.3) above). As discussed in detail in the previous section, for the case of an isolated black hole formed by gravitational collapse, only the particle states which correspond to modes emerging from the "white hole horizon" will be thermally populated according to eq. (7.2.10) at late times. However, since these modes dominate all modes near the black hole horizon, the deviation from an exactly thermal state (with respect to all modes) near the horizon is negligible, and will be ignored in our calculations below. (In any case, the deviations from thermality due to the fact that the particle states in $\mathcal{H}_{-\infty}$ are unpopulated should serve only to enhance the effects calculated below.)

Since the redshift factor χ is not constant, according to eq. (7.2.10) there will be a nonzero gradient of the locally measured temperature seen by static observers. By the Gibbs-Duhem relation of elementary thermodynamics (in the case of vanishing chemical potential) there will be a pressure gradient associated with the thermal bath given by

$$\nabla_a P = s \nabla_a T \tag{7.2.11}$$

where s is the entropy density of the thermal bath. Consequently, there will be a buoyancy force exerted on a box lowered slowly toward a black hole, much as though the box were being lowered into an ordinary fluid body. I will now outline the calculation (Unruh and Wald 1982) that shows that this buoyancy force acts in just the right way to preserve the generalized second law.

Taking into account the buoyancy force, we find that the force, eq. (7.2.6), is modified to become

$$F_\infty = E\frac{d\chi}{dy} + V\frac{d(\chi P)}{dy} \qquad (7.2.12)$$

where V denotes the volume of the box. Integrating eq. (7.2.12), we find

$$W_\infty = (1-\chi)\,E - \chi PV \qquad (7.2.13)$$

so that the energy delivered to the black hole now is given by

$$\Delta M = \chi\,(E + PV) \qquad (7.2.14)$$

Thus, more energy actually is delivered to the black hole than was found in the above classical calculation. Indeed, since χP becomes large near the horizon, the optimal place to release the matter into the black hole no longer is at the black hole horizon. Rather, the optimal place now occurs at the value of y at which

$$0 = \frac{d(\Delta M)}{dy} = -\frac{dW_\infty}{dy} = -F_\infty \qquad (7.2.15)$$

i.e., at the "floating point" of the box. By eq. (7.2.12), the "floating point" condition is

$$\begin{aligned} 0 &= E\frac{d\chi}{dy} + PV\frac{d\chi}{dy} + V\chi\frac{dP}{dy} \\ &= (E+PV)\frac{d\chi}{dy} + V\chi s\frac{dT}{dy} \\ &= (E+PV-VsT)\frac{d\chi}{dy} \end{aligned} \qquad (7.2.16)$$

where eqs. (7.2.10) and (7.2.11) were used in the last two steps. Since $d\chi/dy \neq 0$, the floating point condition becomes

$$E + PV - sTV = 0 \tag{7.2.17}$$

However, the integrated form of the Gibbs-Duhem relation for the thermal bath yields

$$eV + PV - sTV = 0 \tag{7.2.18}$$

where e denotes the energy density of the thermal bath. Thus, the condition for the box to float is simply

$$E = eV \tag{7.2.19}$$

which agrees with a result previously obtained by Archimedes.

We substitute eq. (7.2.19) into eq. (7.2.14) to obtain the minimum energy that can be delivered to the black hole in this process. Using eqs. (7.2.18) and (7.2.10), we obtain

$$(\Delta M)_{min} = \frac{\kappa}{2\pi} Vs \tag{7.2.20}$$

and hence,

$$(\Delta A)_{min} = \frac{8\pi}{\kappa} (\Delta M)_{min} = 4Vs \tag{7.2.21}$$

Thus, the net change in generalized entropy in the process is given by

$$\Delta S' = \Delta S_m + \Delta A/4 \geq \Delta S_m + (\Delta A)_{min}/4 = -S + sV \tag{7.2.22}$$

where s is the entropy density of the thermal bath at the floating point. But, by definition, at a given energy and volume, the entropy is maximum in a thermal state. Thus, taking eq. (7.2.19) into account, we obtain $sV \geq S$ and thus,

$$\Delta S' \geq 0 \tag{7.2.23}$$

i.e., the generalized second law cannot be violated by this process. Note that this argument does not assume the existence of any bound (such as the one postulated by Bekenstein (1981)) for the amount of entropy at fixed energy that can be put in a box of a given size. If S/E is large for the matter lowered toward the black hole, then the entropy density of thermal matter at that energy density must be correspondingly large, and, by eq. (7.2.11), the buoyancy force on the box also will be correspondingly large, thereby preventing a violation of the generalized second law.

Note also that in the above calculation of the buoyancy force on the box, we attributed an energy density, e, and pressure, P, to the thermal bath of radiation. In fact, however, this is not correct: For a macroscopic black hole, the true expected stress-tensor $<T_{ab}>$ of the quantum field is negligibly small near the horizon, as expected on physical and dimensional grounds. The thermal bath values, e and P, used in the above calculation actually measure the expected energy and pressure relative to the natural vacuum state, $|0>_s$, defined by the static isometries. (Thus, it follows that for a macroscopic black hole, the expected energy density and pressure in the state $|0>_s$ are nearly $-e$ and $-P$, respectively.) Since only the stress-tensor differences between the outside and inside of the box are relevant to the calculation of forces on the box, this shift of the "zero-point" of $<T_{ab}>$ has no effect upon the above results. However, it does indicate that the process is more accurately described by saying that—rather than feeling an externally applied buoyancy force—the box fills up with negative energy and pressure (flowing into the box by a "radiation by moving mirrors" effect) as it is slowly lowered. In this more accurate "inertial" description, the "floating point" occurs when a sufficient amount of negative energy has flowed into the box that the total energy of the box is zero. The difference between the behavior of a slowly lowered box—which feels a large force of quantum origin—and a freely falling box—which feels a negligible quantum force—also is readily explained in this inertial viewpoint, since the freely falling box does not fill up with negative energy. Further discussion of these features of the process as well as a discussion of the inverse process of "mining" energy from a black hole can be found in Unruh and Wald (1982).

The fact that the above gedankenexperiment fails to produce a counterexample to the generalized second law suggests that it should be possible to give a more general argument for the validity of that law, at least in the case of processes which can be treated as small perturbations of a stationary black hole. Such an argument has been given by Zurek and Thorne (1986) and Thorne, Zurek, and Price (1986). The following argument (Wald 1988, 1992) roughly corresponds to the one they have given; a related argument has been given very recently by Frolov and Page (1993).

We consider a process wherein we start with a stationary black hole and perturb it (infinitesimally) by some process, e.g., by dropping matter into it. We wish to calculate the net change, $\Delta S'$, in generalized entropy resulting from the process. In comparing the perturbed spacetime with the unperturbed black hole, it is convenient to make an identification of perturbed and unperturbed spacetimes in such a way that the black hole horizons coincide and have the same null generators. In addition, we identify the spacetimes so that in a neighborhood of the horizon of the perturbed spacetime, the image under this identification of the Killing field χ^a normal to the horizon in the unperturbed spacetime has the same norm, $\chi^2 = -\chi_a\chi^a$, as it has in the unperturbed spacetime. (This can be achieved by composition of any horizon preserving identification with an additional diffeomorphism which moves points along the orbits of χ^a, thereby "compressing" or "stretching" χ^a as needed.) We then define χ^a on the perturbed spacetime to be the image of χ^a under this identification of the Killing field χ^a. Thus, in this choice of "gauge", we automatically have $\delta\chi^a = 0$ on the perturbed spacetime as well as $\delta\chi = 0$ in a neighborhood of the horizon.

Consider, now, the family of observers outside the black hole who follow orbits of χ^a. In the unperturbed spacetime, such observers "see" a thermal bath of particles, and—relative to the stationary vacuum state $|0>_s$ associated with χ^a—they would assign a thermal bath energy density, e, to the quantum field given by

$$e = T_{ab}\chi^a \chi^b/\chi^2 \tag{7.2.24}$$

where T_{ab} denotes the difference between the actual expected stress energy and the expected stress-energy in the state $|0>_s$. Such observers also would naturally assign to the quantum field a thermal bath entropy current of the form

$$S^a = s\chi^a/\chi \tag{7.2.25}$$

However, as I shall discuss further below, in this case the "objective significance" of S^a is far from clear. The local entropy density, s, then is given in terms of S^a by

$$s = -S_a \chi^a/\chi \tag{7.2.26}$$

Consider, now, the perturbed spacetime and consider, again, the observers following orbits of χ^a. The perturbation in the energy and entropy densities they would assign to the quantum field are given by

$$\delta e = \delta[T_{ab}\, \chi^a \chi^b/\chi^2]$$

$$= (\delta T_{ab}) \chi^a \chi^b/\chi^2 \qquad (7.2.27)$$

and

$$\delta s = -\delta[S_a\, \chi^a/\chi] = -(\delta S_a)\, \chi^a/\chi \qquad (7.2.28)$$

However, δs would be maximized for a given δe if the perturbed field remained locally in a thermal state. Hence, we must have

$$\delta s \leqq (\delta s)_{th} = \delta e/T = 2\pi\chi\, \delta e/\kappa \qquad (7.2.29)$$

where we have used the ordinary first law of thermodynamics for the thermal bath as well as the formula (7.2.10) for the locally measured temperature. We multiply eq. (7.2.29) by χ and take the limit as one approaches the horizon, using eqs. (7.2.27) and (7.2.28). We thereby obtain

$$-(\delta S_a)\, \chi^a\Big|_{\text{horizon}} \leqq \frac{2\pi}{\kappa}\, (\delta T_{ab})\, \chi^a \chi^b\Big|_{\text{horizon}} \qquad (7.2.30)$$

We integrate eq. (7.2.30) over the horizon (with respect to Killing parameter v). The left side then can be interpreted as the total flux of matter entropy into the black hole, whereas the right side is proportional to the same combination of energy and angular momentum fluxes as appeared in the derivation of the first law (see eq. (6.2.6)). (Although T_{ab} is the "fictitious thermal bath stress-energy", the difference δT_{ab} is equal to the difference in true stress energies.) Using the first law of black hole mechanics (6.2.8), we thus obtain the final result

$$-\Delta S \leqq \Delta A/4 \qquad (7.2.31)$$

which states that the generalized second law is satisfied in any process that can be treated as a small perturbation of a stationary black hole.

Although the above calculation certainly is supportive of the validity of the generalized second law, we should point out a problematical aspect of the analysis. As already noted above, the calculation assumes that we can assign an entropy current density S^a to the state of the quantum field in such a way that eqs. (7.2.26) and (7.2.29) hold and $-S_a\chi^a$ yields the matter entropy flux into the black hole. However, it is far from clear how to define a meaningful notion of an entropy current density in quantum field theory, particularly since one would wish to avoid the introduction of the problematical notion of "particles" in making such a definition. Indeed, the likely difficulties in defining S^a in such a way as to validate the above argument are highlighted if we consider the entropy current to be assigned to the Minkowski vacuum state in flat spacetime. By Poincare invariance, it is clear that any "objectively defined" entropy current density must vanish. However, a uniformly accelerating observer following the orbit of a boost Killing field χ^a would naturally assign a nonvanishing entropy current of the form (7.2.25) to this state, and, indeed, this type of assignment was assumed in the above calculation. One might expect that—in analogy with the situation for stress-energy—formula (7.2.25) actually should be interpreted as yielding the difference between a well defined, "objective" entropy current density of the Minkowski vacuum state (which vanishes) and that of the static (i.e., "Rindler") vacuum associated with χ^a. However, that would entail the assignment of a negative entropy density to Rindler vacuum state in Minkowski spacetime, and it is not clear what this would mean. Alternatively, perhaps one should not attempt to define an "objective" entropy, but only define entropy with respect to a notion of time translations, so that, for example, the entropy current of the Minkowski vacuum with respect to ordinary time translations would vanish, but its entropy with respect to χ^a would be nonvanishing. However, in that case, it would appear that entropy could be defined only in the quasi-stationary case.

Despite this difficulty, I see little reason to doubt the validity of the generalized second law. (In any case, the recent arguments of Frolov and Page (1993) do not require the introduction of a local entropy current density.) If the validity of the generalized second law is accepted, then by far the most natural interpretation of the laws of black hole thermodynamics is that they simply *are* the ordinary laws of thermodynamics applied to a black hole. In that case, A/4 truly would represent the physical entropy of a black hole, and S' simply would be the total entropy of the universe, including contributions both from ordinary matter and from black holes. Indeed, in the absence of a complete quantum theory of gravity, it is hard to imagine how a more convincing case could be made for this conclusion.

Nevertheless, some major puzzles remain. Prominent among them are the following two: (1) Underlying ordinary thermodynamics and the usual interpretation of entropy is the idea that "time average = phase average" for general dynamical systems. In view of the nature of "time" in general relativity, it is hard to see how this notion would be applicable to a system containing a black hole, and, if it is not, what idea would replace it. In addition, the fact that a black hole cannot causally influence its exterior makes it difficult to understand the underlying mechanism by which thermal equilibrium could be achieved between a black hole and a material body. (2) Why is the entropy of a black hole so simply and directly related to its horizon area, even in the nonequilibrium case? A formula of the desired type could arise if all the degrees of freedom of a black hole were concentrated in a Planck length "skin" around the horizon. Namely, if a finite number of states are assigned to each Planck volume in this region, then the logarithm of the total number of states would be proportional to A. Some ideas for how to derive such a formula for entropy have been proposed by Sorkin and collaborators (see Bombelli et al. (1986), and references cited therein; see also 't Hooft (1991, 1992)). However, these ideas run counter to the notion in classical general relativity of the black hole horizon as being a globally defined, mathematical surface, possessing no local, physical significance, and thus providing a very poor candidate for where the true dynamical degrees of freedom of a black hole should lie.

Further investigation of these issues may well represent our best opportunity to gain further insight into the nature of quantum gravity.

7.3 Evaporation of Black Holes and Loss of Quantum Coherence

Another striking ramification of the Hawking effect arises from consideration of "back-reaction" effects. In the (fixed, background) space-time of fig. 7.1, the energy properties of the quantum field are described semiclassically by its expected stress-energy tensor $<T_{ab}>$ (see section 4.6). Since the spacetime is static outside of the collapsing matter, the expected energy current

$$J_a = \langle T_{ab} \rangle t^b \tag{7.3.1}$$

is conserved in that region, where t^b denotes the static Killing field of the Schwarzschild geometry. However, the particle creation calculation shows that at infinity—where the spacetime is nearly flat and, hence, normal ordering with respect to the static vacuum can be used to calculate $\langle T_{ab} \rangle$—there will be a steady nonzero flux, F, associated with J^a. On dimensional grounds, for a massless field this flux must be of the form

$$F = \frac{\alpha}{M^2} \tag{7.3.2}$$

where α is of order unity in Planck units. The value of α can be estimated from the Stefan-Boltzmann flux law, $F_{S-B} = \sigma A T^4$, with $T = 1/(8\pi M)$ and $A = 4\pi(3\sqrt{3}M)^2$, where $3\sqrt{3}M$ is the effective radius of the black hole when "light bending" is taken into account. However, since the modes predominantly contributing to the flux have wavelength comparable to the size of the black hole, the Stefan-Boltzmann law holds only approximately, and a ("physical optics") mode sum must be done to compute α exactly. Note that eq. (7.3.2) also will hold approximately for any field whose mass is small compared with the black hole temperature (7.1.9), but the flux for fields whose mass is much greater than the black hole temperature would be negligible.

By conservation of J^a, a negative energy flux equal in magnitude to (7.3.2) must flow into the black hole. Consequently, it is clear that back-reaction effects will cause the black hole to lose mass. In order to calculate these back-reaction effects in the semiclassical approximation, we would like to self-consistently solve the semiclassical Einstein equation,

$$G_{ab} = 8\pi \langle T_{ab} \rangle \tag{7.3.3}$$

Unfortunately, for the three reasons discussed near the end of section 4.6, this has not been done. Nevertheless, on physical grounds, one would expect that for a black hole whose mass, M, is much greater than the Planck mass, $M_P = (c\hbar/G)^{1/2} \approx 10^{-5}$ gm, the back-reaction effects will be locally small near the black hole horizon and in the entire exterior region (i.e., everywhere except very near the singularity within the black hole). In that case, the spacetime geometry should be accurately represented by a locally Schwarzschild geometry in which M decreases slowly with time in accordance with eq. (7.3.2). Thus, considering, for simplicity, only the contribution from massless fields, we obtain

$$\frac{dM}{dt} = -F = -\frac{\alpha}{M^2} \tag{7.3.4}$$

which is easily integrated to yield

$$M(t) = (M_0^3 - 3\alpha t)^{1/3} \tag{7.3.5}$$

Thus, we find that a black hole should completely "evaporate" (or, at least, reach the Planck mass, at which point this calculation certainly no longer can be trusted) in the finite time $M_0^3/3\alpha$.

A spacetime diagram depicting the process of black hole formation and evaporation is shown in fig. 7.2, under the assumption that complete evaporation occurs. The global manner in which M decreases—i.e., the region of spacetime corresponding to a particular value of M—is indicated in this figure. It should be noted (Wald 1986) that the spacetime of fig. 7.2 contrasts sharply with that of a Vaidya solution with the same mass dependence at future null infinity. In the Vaidya solution one has an everywhere positive energy flux which is locally large near the horizon, whereas in the spacetime of fig. 7.2, the energy flux is locally negative near the horizon and is locally small everywhere outside of the black hole (until the black hole reaches Planck dimensions).

One immediate, noteworthy ramification of the idea the black holes can "evaporate" due to the Hawking effect is that conservation laws for lepton and baryon number apparently can be grossly violated. Indeed, one could imagine forming a black hole with $M \sim M_\odot$ by the collapse of a neutron star, with baryon number $\sim 10^{57}$. This baryon number should not leave any "imprint" upon the black hole exterior at late times, so when the black hole evaporates, it should do so in a baryon-antibaryon symmetric manner. If so, then the expected baryon number after black hole evaporation would be zero. Indeed, one would expect, in any case, that essentially all of the energy of the black hole would be radiated in particles which carry zero baryon number, since the black hole temperature is much less

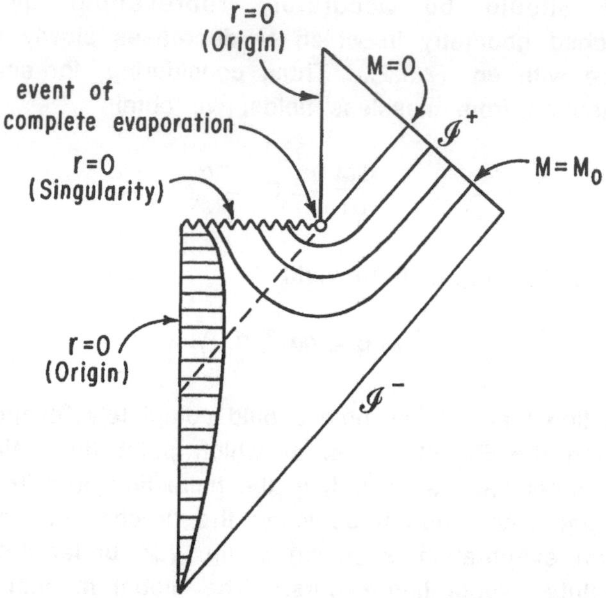

Figure 7.2. A conformal diagram of a spacetime in which black hole formation and evaporation occurs. Below the contour labeled "M=M_0" the region exterior to the collapsing body is well approximated by the Schwarschild geometry of the initial mass M_0. In a neighborhood of each of the other contours, the spacetime geometry also is well approximated by the Schwarzschild metric, but with a smaller value of mass. The contour labeled "M=0" lies at the retarded time corresponding to the final instant of black hole evaporation. If, for simplicity, we assume that all of the Hawking radiation emerges in the form of massless particles, then the spacetime geometry will be flat above this contour.

than the mass of any baryon until the very late stages of evaporation. Thus, in order to have the possibility of conserving baryon number in the process of black hole formation and evaporation, it would be necessary to find a mechanism which would "shut off" the Hawking radiation of photons, neutrinos, etc., with essentially 100% efficiency. It seems highly implausible that such a mechanism exists.

The idea that black holes can "evaporate" gives rise to another striking ramification of the Hawking effect: the "loss of quantum coherence." To explain this phenomenon, consider, first, the behavior of the quantum field in the spacetime of fig. 7.1, in which back-reaction effects are not taken into account. Our analysis in section 7.1 showed that the state of the field at late times in region I—and, in particular, the particle flux reaching infinity—must be described by a density matrix. In the S-matrix analysis, this occurred because the particles present in region I at late times are strongly correlated with particles which entered the black hole at early times. (As seen from our analysis, these correlations are closely analogous to the correlations occurring between particles in region I and II of fig. 5.1 in the Unruh effect, as described in section 5.1.) In the algebraic viewpoint, this effect may be readily understood in the following manner. Consider a Cauchy surface for the spacetime of fig. 7.1 (such as the Cauchy surface Σ shown there) which intersects the black hole event horizon. Break up Σ into the portions Σ_I and Σ_{III}, which, respectively, lie outside and inside the black hole. Consider the globally hyperbolic spacetime regions $\text{int}\, D(\Sigma_I)$ and $\text{int}\, D(\Sigma_{III})$, where D denotes the domain of dependence (see eq. (4.1.3)) and "int" denotes the interior of a set. (Note that $I^+(\Sigma_I) \cap \text{int}\, D(\Sigma_I)$ consists precisely of all of region I to the chronological future of Σ.) Then the fact that in the S-matrix calculation a density matrix must be used to describe the state of the field in region I at late times corresponds simply to the fact that the restriction of the state of the field to the subalgebra associated with the region $\text{int}\, D(\Sigma_I)$ yields a mixed state. As discussed near the end of section 4.5, this is to be expected since the domain of determinacy of $\text{int}\, D(\Sigma_I)$ is not the entire spacetime. Even if the state on the entire Weyl algebra is taken to be pure, correlations of the field should occur between the regions $\text{int}\, D(\Sigma_I)$ and $\text{int}\, D(\Sigma_{III})$—i.e., between the regions outside and inside the black hole at "time" Σ.

Consider, now, the spacetime of fig. 7.3, in which back-reaction causes the black hole to "evaporate." In the S-matrix approach, the particles propagating out to infinity again will be described by a density matrix for the same reason as in the spacetime of fig. 7.1: The particles propagating to infinity are correlated with particles which enter the black hole. As in the previous case, particle creation and scattering will be described by an ordinary S-matrix, provided that the particles which propagate into the black hole are represented in the "out" Hilbert space. However, an important difference occurs in the spacetime of fig. 7.3: The black hole disappears from the spacetime, so at late times, the "entire" state of the field is mixed. Thus, if one takes the "out"

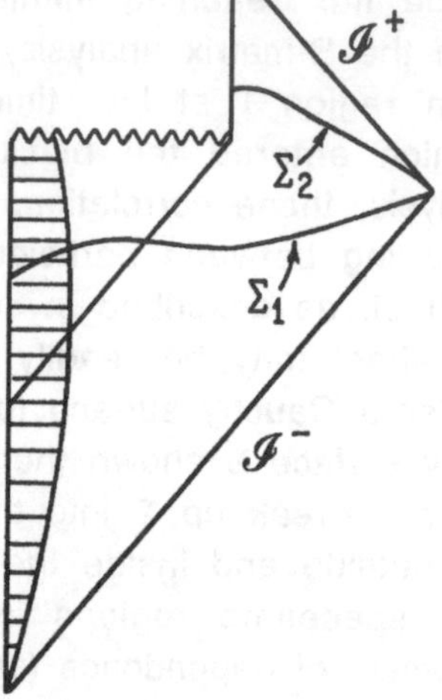

Figure 7.3. A conformal diagram illustrating the phenomenon of loss of quantum coherence in a spacetime in which black hole evaporation occurs.

Hilbert space to be the Fock space associated with particles propagating out to infinity at late times, one cannot describe particle creation and scattering by an ordinary S-matrix, since an initial pure state will evolve to a final density matrix. (However, a "superscattering matrix" can be used to describe this scattering; see Hawking (1976) and Wald (1980) for further discussion.) This is the phenomenon of "loss of quantum coherence."

In the algebraic viewpoint, this phenomenon can be understood as follows. Consider the hypersurface, Σ_2, of fig. 7.3, corresponding to a "time" after which the black hole has evaporated. Then $\text{int}\, D(\Sigma_2)$ includes the "entire future" of the spacetime. However, the domain of determinacy, $\mathscr{D}(\text{int}\, D(\Sigma_2))$, of $\text{int}\, D(\Sigma_2)$ fails to be the entire spacetime, since, in particular, it does not include any portion of the black hole region. Hence, as discussed above, it should not be surprising that the restriction of a pure state, ω, to the subalgebra associated with the region $\text{int}\, D(\Sigma_2)$ should be mixed, since correlations of the field will occur between this region and the interior of the black hole. On the other hand, the domain of determinacy of the hypersurface Σ_1 plausibly will be the entire spacetime (see Wald

(1984b)). In that case, the "evolution" of the state from "time" Σ_1 to "time" Σ_2 will correspond to the restriction of the state from the Weyl algebra of the entire spacetime to the Weyl algebra for int $D(\Sigma_2)$. Hence, one will obtain evolution from a pure state to a mixed state.

This evolution from a pure state to a mixed state is commonly regarded as corresponding to a serious breakdown of quantum theory. I do not share this view. Exactly the same sort of phenomenon occurs for a massless quantum field in flat spacetime if one considers its evolution from an initial Cauchy hypersurface, Σ_1, (such as a hyperplane) to a final hypersurface, Σ_2, (such as a hyperboloid) which fails to be a Cauchy surface for Minkowski spacetime (see fig. 7.4). In this case, the field in int $D(\Sigma_2)$ will be correlated with radiation propagating out to infinity rather than radiation propagating into a black hole but the consequence is the same: the final state will be mixed even if the initial state is pure. Thus, again there is a "loss of quantum coherence." Scattering cannot be described using an ordinary S-matrix, and a superscattering matrix must be employed. In both this case and the case of the evaporating black hole, this phenomenon is entirely attributable to the failure of the "final" hypersurface, Σ_2, to be a Cauchy surface, so that the "final state" does not provide a complete description of the "full state" of the field. Thus, in both cases the "breakdown" of ordinary S-matrix scattering is directly attributable to the incompleteness of the characterization

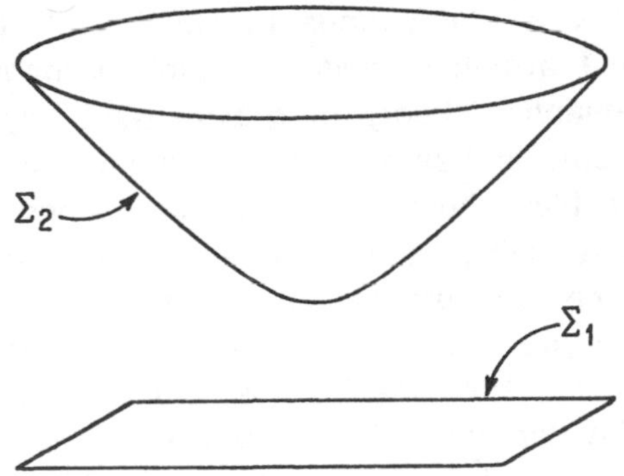

Figure 7.4. A spacetime diagram showing a hyperplane, Σ_1, and a hyperboloid, Σ_2, in Minkowski spacetime. Evolution from Σ_1 to Σ_2 results in a loss of quantum coherence similar to that occurring in fig. 7.3.

of the field in terms of its final state. It should be emphasized that there is no corresponding "breakdown" in any of the *local* laws governing the behavior of the quantum field. In particular, in both cases all local conservation laws will continue to hold and no qualitatively new phenomena will be seen by any family of observers making measurements over a compact region of spacetime.

Nevertheless, although the above semiclassical description of the phenomenon of loss of quantum coherence in black hole formation and evaporation does not correspond to any breakdown in the ordinary local dynamical laws of quantum field theory, it does suggest the possibility that these local laws might be significantly modified when quantum gravity itself is taken into account. Namely, if we grant that, under appropriate circumstances, "real" black holes can form and evaporate in quantum gravity, one might expect that, under all circumstances, "virtual" black holes should have a nonzero amplitude to "mediate" processes in which a pure state evolves to a mixed state. In that case, the effective, "macroscopic" (compared with the Planck scale) local dynamical laws for a quantum field might well yield a nonzero probability for evolution from a pure to mixed state. However, there appear to be some serious difficulties with formulating causal and energy conserving local evolution laws in which pure states locally evolve to mixed states (Banks, Susskind, and Peskin 1984). Thus, in this sense, there appears to be a potential conflict between the the phenomenon of black hole evaporation and the existence of acceptable local evolution laws in ordinary quantum field theory.

My own view is that serious potential conflicts with the usual properties of local evolution laws in quantum theory are inherently present in any quantum theory of gravity (i.e., any theory in which the spacetime metric is treated as a quantum field), whether or not the theory admits black holes. In particular, if the spacetime metric is not treated classically, there should be "fluctuations" possible in the local light cone structure (which, presumably, would be large at the Planck scale), so the causal relationships between events will not be well defined. Presumably, this should give rise to a (small) nonzero probability for acausal propagation in the effective, "macroscopic" local dynamical laws in quantum field theory. The key point is that, at present, one can at best make guesses about the true, fundamental character of dynamical evolution in a quantum theory of gravity, and any proposed quantum theory of gravity will face major challenges in attempting to account for how the ordinary, local, dynamical laws of quantum field theory in a classical spacetime emerge from the fundamental dynamics. The inclusion of phenomena in quantum gravity corresponding to black hole formation and evaporation does not, in my view, significantly enhance this challenge.

Nevertheless, the loss of quantum coherence predicted to occur in the process of black hole formation and evaporation by the above semiclassical arguments is a remarkable phenomenon—which, indeed, might provide some important clues to the fundamental nature of dynamics in quantum gravity—so it is of interest to critically examine the assumptions which lead to this prediction to see if there exists a plausible means by which the conclusion might be evaded. If we exclude radical proposals which introduce new physics well before the spacetime curvature reaches the Planck scale, there appear to be only two basic alternative models of the evaporation process in which "loss of quantum coherence" does not occur:

(1) A Planck mass "remnant" of the black hole remains which has sufficiently many states to contain all of the "correlated information" of the evaporation process. The field outside the remnant is described by a highly mixed state as above, but the total state of the field remains pure.

(2) The correlations between particles emitted at early times and the quantum state of the black hole are somehow transferred, as the black hole evaporates, to particles which are emitted at later times. This could occur in either of the following two ways: (a) Most of the correlated information emerges when the black hole has reached Planck dimensions. (b) The correlated information emerges gradually, as the black hole slowly evaporates. In either of these subcases, the black hole evaporates completely in the manner described above, but the final state of the field is pure.

In my view, possibility (1) is highly implausible, since the initial black hole could be arbitrarily large, so the Planck mass remnant would have to admit arbitrarily many internal states in order to correlate with the emitted particles. The arbitrarily large number of internal states thereby required would seem to be in sharp conflict with the formula (7.2.1) for black hole entropy, which suggests that a black hole of Planck dimensions should have ~1 internal state. In addition, unless the remnant and its internal states are observable, possiblity (1) actually would not be physically distinguishable from a model where true loss of quantum coherence occurs.

Possiblity (2a) also suffers from the fact that it would require a black hole of Planck dimensions to have an arbitrarily large number of internal states. In addition, this black hole would be required to emit matter in a state of enormous entropy, but with only $\sim M_P \sim 10^{-5}$ gm total energy. In order to do this, it presumably would be necessary to distribute the energy in the form of massless particles of arbitrarily long wavelength. It seems difficult to imagine a mechanism by which an object of size $\sim l_P \sim 10^{-33}$ cm could efficiently emit particles of arbitrarily long wavelength.

Possibility (2b) probably is the most plausible of the alternative possibilities. Indeed, this is the manner in which an ordinary material body—initially in a pure state—would cool to $T = 0$ by emission of photons: The photons emitted at early times are correlated with the atoms which emitted them, and these correlations are then gradually transferred to the photons emitted at later times. However, there is a crucial difference between an evaporating black hole and an ordinary hot body. When the black hole is much larger than the Planck size, a classical description of the spacetime geometry outside of the black hole should be an excellent approximation. However, in that case, information about the state of the field inside the black hole cannot causally propagate outside the black hole. Thus, unlike the case of a hot body, it is difficult to imagine a mechanism so efficient that it would restore all of the correlations and yet not produce a gross violation of causality.

Indeed, the difficulties which arise with arguments along the lines of possibility (2b) are greatly magnified if one considers the evaporation process for a large black hole formed by the collapse of a subsystem of quantum matter which is in a highly mixed state, but is correlated with distant matter in the universe, so that the initial state of the complete system is pure. For example, suppose that we start with a pure state of matter corresponding to a large collection of pairs of spin–1/2 particles, each of which is in a spin–0 state, so that the spin of the first particle in each pair is perfectly anti-correlated with the spin of the second. We now build a star of mass M out of the collection of "first particles" in each pair, and disperse the "second particles" to distant regions of the universe. We then allow this star to collapse to a black hole. For sufficiently large M, this black hole will form at a stage in which all the usual local measures of the importance of quantum effects (energy density of matter, spacetime curvature, etc.) are negligibly small. The matter will rapidly plunge into the deep interior of the black hole (where, classically, it is predicted to disappear into a spacetime singularity). In order for the black hole to evaporate in such a way that the final state is pure, it is necessary that the emitted Hawking radiation be strongly correlated with the spin states of the distant matter. In order for this to be possible, it seems necessary that either the collapsing matter must exert a (highly acausal) influence on the particle creation process from deep inside the black hole, or that it somehow must leave an imprint on the spacetime structure exterior to the black hole which is so powerful that it affects the particle creation process in such a way as to restore all of the initial correlations with the distant matter. It seems very difficult to imagine a mechanism by which either of these alternatives could occur with the efficiency needed to produce a final pure state—or, if there were such a mechanism, why its effects would not easily be seen in (nearly) flat spacetime physics.

Thus, in my view, by far the most plausible outcome of the particle creation process is that the black hole evaporates completely and that an initially pure state evolves to mixed state. Nevertheless, a completely definitive determination of the behavior of an evaporating black hole will not be available until a complete calculation—taking into account all back-reaction and quantum effects—can be done. Such a complete calculation would require, of course, a complete formulation of a quantum theory of gravity, so there is little prospect that such a calculation can be undertaken in the foreseeable future. However, a semiclassical back-reaction calculation within the framework described at the end of section 4.6 could, at least, provide an important first step toward such a calculation, and should provide a strong indication as to whether the basic picture of the evaporation process described above is correct. Unfortunately, for the reasons discussed at the end of section 4.6, even such a semiclassical analysis does not appear to be a feasible undertaking at the present time.

A natural approach to dealing with this situation is to consider a simpler "toy theory" in which at least the semiclassical analysis could be undertaken. Ordinary general relativity in fewer than four dimensions is dynamically trivial, and thus is not a suitable candidate for such a theory. However, the addition of a "dilaton" scalar field to lower dimensional general relativity results in a theory which is dynamically nontrivial, but still appears to mimic many of the features of four-dimensional general relativity. Recently, investigations of black hole evaporation have been undertaken (Callen et al. 1992) on a field theory model in two-spacetime-dimensions in which the dynamical variables consist of a Lorentz metric, g_{ab}, a dilaton scalar field, ψ, and an ordinary scalar field, ϕ. The ("string inspired") action of the theory is taken as

$$S = \frac{1}{2\pi} \int d^2x \sqrt{-g} \left\{ e^{-2\psi}[R + 4\nabla_a\psi\nabla^a\psi + 4\Lambda^2] - \frac{1}{2} \nabla_a\phi\nabla^a\phi \right\} \qquad (7.3.6)$$

where Λ is a "cosmological constant". Classically, the field equations derived from this action admit solutions with spacetime structure analogous to fig. 7.1, which describe the gravitational collapse of the matter field ϕ to form a black hole. Furthermore, classically, there exist laws of black hole mechanics in this theory which are completely analogous to the general relativistic laws discussed in section 6.2 (Frolov 1992). Thus, this theory would appear to provide a suitable testing ground for ideas on the quantum behavior of black holes. Of course, there is no guarantee that the semiclassical (or fully quantum) behavior of this or similar models will yield an accurate description of the four-dimensional theory.

If one treats ϕ as a (linear) quantum field propagating in a classical, background spacetime corresponding to the formation of a black hole by gravitational collapse, then the same analysis as given in section 7.1 for ordinary general relativity shows that Hawking radiation will result. One may then investigate the semiclassical back-reaction effects of the Hawking radiation in this model by dynamically evolving the classical fields, g_{ab} and ψ, and the quantum field ϕ via the analog for this model of the semiclassical Einstein equation (7.3.3). Remarkably, the three formidable difficulties of the semiclassical Einstein equation in four dimensions discussed in section 4.6 either do not occur or are significantly alleviated in this model.

First, in two dimensions there is no analogous "local curvature ambiguity" in the expected stress energy, $\langle T_{ab} \rangle$, of the scalar field. This can be understood from the fact that in two dimensions, $\langle T_{ab} \rangle$ should have dimension (length)$^{-2}$, so the appropriate "local curvature terms" would have to be linear in the curvature. However, since the Einstein tensor vanishes identically in two dimensions, no conserved local curvature terms of the correct dimension exist.

Secondly, for essentially the same reason, $<T_{ab}>$ contains derivatives of the metric only up to second (rather than fourth) order, and thus does not drastically alter the second order differential character of the classical equations. Hence, the semiclassical evolution equations of this model are better behaved mathematically than their four-dimensional counterparts. In particular, the equations should not admit any new, spurious, "runaway" solutions.

Third, for the case occurring in this model of a massless Klein-Gordon field—which is conformally invariant in two dimensions—$<T_{ab}>$ is easily computed in a general curved background. (This computation is greatly facilitated by the fact that an arbitrary curved spacetime is locally conformally flat in two dimensions.) Furthermore, for the "in" vacuum state, $<T_{ab}>$ is given by a simple formula which is local in the conformal factor (see, e.g., Wald (1978b) for further details). Thus, in this model, the semiclassical evolution equations are simply local partial differential equations of the same differential order as in the classical theory.

However, it turns out that the new terms in the semiclassical equations arising from $<T_{ab}>$ do change the character of the evolution equations sufficiently that situations can arise where the fields are nonsingular but their dynamical evolution is not well defined. Nevertheless, this difficulty can be treated by the imposition of appropriate boundary conditions, and in addition, modifications of the original model (7.3.6) also can be considered which make it possible to solve the semiclassical equations analytically (Russo et al. 1992). The spacetime structure and quantum field behavior of the solutions thus obtained yield a picture of black hole evaporation consistent with the above discussion in the sense that it is found that the black hole radiates most of its mass in Hawking radiation (at which point the semiclassical approximation should break down), and no restoration of correlations occurs by this stage. Of course, if the restoration of correlations were to occur, one would not necessarily expect it to occur within the context of the semiclassical approximation, so it would be of considerable interest to study these "toy models" beyond the semiclassical approximation. Such investigations are presently being pursued, and one may hope that they will yield some further insights into the nature of black hole evaporation.

Appendix: Some Basic Definitions and Constructions Pertaining to Hilbert Spaces

The main purpose of this Appendix is to review some basic definitions and constructions relevant to Hilbert spaces and to introduce an index notation, which is very useful for representing operations in Fock space in a basis independent manner. The definition of a C*-algebra also will be given.

A.1 Some Basic Definitions

First, we recall that a vector space, V, over the complex numbers, $\mathbb{C}$, is a set on which is defined an operation of "addition" (which is a map $+: V \times V \to V$) and an operation of "scalar multiplication" (which is a map $\times: \mathbb{C} \times V \to V$) such that (1) V is an abelian group with respect to addition, with identity element denoted $\underline{0}$. (The bar under $\underline{0}$ will be dropped when there is no danger of confusion with $0 \in \mathbb{C}$.) (2) The scalar multiplication and addition operations satisfy

$$c_1 \times (c_2 \times \psi) = (c_1 c_2) \times \psi \tag{A.1.1}$$

$$c \times (\psi_1 + \psi_2) = c \times \psi_1 + c \times \psi_2 \tag{A.1.2}$$

$$(c_1 + c_2) \times \psi = c_1 \times \psi + c_2 \times \psi \tag{A.1.3}$$

$$1 \times \psi = \psi \tag{A.1.4}$$

for all $c, c_1, c_2 \in \mathbb{C}$ and $\psi, \psi_1, \psi_2 \in V$. (Hereafter, the "$\times$" generally will be dropped.)

If V and W are vector spaces, a map $f: V \to W$ is said to be *linear* if $f(c_1\psi_1 + c_2\psi_2) = c_1 f(\psi_1) + c_2 f(\psi_2)$ for all $c_1, c_2 \in C$ and $\psi_1, \psi_2 \in V$. Similarly, f is said to be *antilinear* if $f(c_1\psi_1 + c_2\psi_2) = \bar{c}_1 f(\psi_1) + \bar{c}_2 f(\psi_2)$, where the bar denotes complex conjugation on $\mathbb{C}$.

A *normed vector space* is a vector space, V, together with a map $\| \, \| : V \to \mathbb{R}$ (called a *norm*) satisfying $\|\psi\| > 0$ for all $\psi \neq 0$ and

$$\|c\psi\| = |c|\,\|\psi\| \tag{A.1.5}$$

$$\|\psi_1 + \psi_2\| \leqq \|\psi_1\| + \|\psi_2\| \tag{A.1.6}$$

for all $c \in \mathbb{C}$ and $\psi, \psi_1, \psi_2 \in V$. If V, W are normed vector spaces, a map $f: V \to W$ is said to be ***bounded*** if there exists $C \in \mathbb{R}$ such that $\|f(\psi)\|_W \leqq C\|\psi\|_V$ for all $\psi \in V$.

An *inner product* on a vector space V is a map $<\,,\,> : V \times V \to \mathbb{C}$ which is linear in its second argument and satisfies $\overline{<\psi_1, \psi_2>} = <\psi_2, \psi_1>$ for all $\psi_1, \psi_2 \in V$ (so that $<\,,\,>$ is antilinear in its first argument) and $<\psi, \psi> > 0$ for all $\psi \neq 0$. If $<\,,\,>$ is an inner product on V, then an associated norm on V can be defined by $\|\psi\| = (<\psi, \psi>)^{1/2}$.

If V is a normed vector spaoo, a sequence $\{\psi_n\}$ of elements of V is said to be a *Cauchy sequence* if given $\varepsilon > 0$ there exists $N \in \mathbb{Z}$ such that $\|\psi_n - \psi_m\| < \varepsilon$ for all $n, m > N$. If every Cauchy sequence in V converges then V is said to be *complete* and V is called a *Banach space.* An inner product space which is complete in the associated norm defined above is called a *Hilbert space.*

If V is an inner product space which is not necessarily complete, we can construct the metric space completion, $\mathcal{H}$, of V by taking equivalence classes of Cauchy sequences in V (see, e.g., Reed and Simon (1980)). The inner product space structure of V naturally extends to $\mathcal{H}$ in such a way as to provide $\mathcal{H}$ with the structure of a Hilbert space, with V naturally identified with a dense subspace of $\mathcal{H}$. $\mathcal{H}$ is called the *Hilbert space completion* of V.

An ***algebra***, $\mathcal{A}$, over $\mathbb{C}$ is a vector space over $\mathbb{C}$ with an additional "multiplication map" $\mathcal{A} \times \mathcal{A} \to \mathcal{A}$ which is bilinear (i.e., linear in each variable) and satisfies the associative law. We will denote the action of this map on $(A_1, A_2) \in \mathcal{A} \times \mathcal{A}$ by A_1A_2, so the associative law can be expressed as $A_1(A_2A_3) = (A_1A_2)A_3$. An element $I \in \mathcal{A}$ will be said to be an *identity element* of $\mathcal{A}$ if $AI = IA = A$ for all $A \in \mathcal{A}$.

An algebra, $\mathcal{A}$, will be said to be a **-algebra* if it possesses an antilinear map $*: \mathcal{A} \to \mathcal{A}$ satisfying $A^{**} = A$ and $(A_1A_2)^* = A_2^*A_1^*$ for all $A, A_1, A_2 \in \mathcal{A}$.

An algebra, $\mathcal{A}$, will be called a *normed algebra* if it possesses a vector space norm, $\|\;\|$, (defined above) which, in addition, satisfies $\|A_1A_2\| \leqq \|A_1\|\,\|A_2\|$ for all $A_1, A_2 \in \mathcal{A}$. A normed algebra which also is complete is called a *Banach algebra.*

If an algebra A has the structure of both a *-algebra and a

Banach algebra and if, in addition, we have $\|A^*\| = \|A\|$ for all $A \in \mathcal{A}$, then $\mathcal{A}$ is called a *Banach *-algebra*. Finally, a *C*-algebra*, $\mathcal{A}$, is a Banach *-algebra which also satisfies $\|A^*A\| = \|A\|^2$ for all $A \in \mathcal{A}$.

A.2 Some Basic Constructions

Let $(V,+,\times)$ denote a set, V, together with an "addition map", +, and "scalar multiplication map", $\times$, which satisfy the vector space axioms given above. Define a new "scalar multiplication map" $\bar{\times} : \mathbb{C}\times V \to V$ by

$$c \,\bar{\times}\, \psi = \bar{c} \times \psi \tag{A.2.1}$$

Then it is easily checked that $(V,+,\bar{\times})$ also satisfies the vector space axioms. The resulting vector space is called the *complex conjugate space* to the original vector space. If the original vector space is denoted simply as V, the complex conjugate space usually is denoted as $\overline{V}$. (Note, however, that the notation is potentially confusing since the two vector spaces differ in their scalar multiplication maps rather than their underlying sets. Note, also, that the natural one-to-one, onto correspondence between V and $\overline{V}$—resulting from their being constructed from the same underlying set—is antilinear.) If $\psi \in V$, we denote the vector in $\overline{V}$ associated with ψ via this correspondence by $\overline{\psi}$. If $A: V \to W$ is a linear map, we denote by $\overline{A}$ the corresponding linear map from $\overline{V}$ to $\overline{W}$. Finally, if $<\,,\,>$ is an inner product on V, then an inner product on $\overline{V}$ can be defined by

$$<\overline{\psi}_1, \overline{\psi}_2>_{\overline{V}} = <\psi_2, \psi_1>_V \tag{A.2.2}$$

Let V be a normed vector space. The *dual space* to V—denoted as V*—is defined as the set of all bounded linear maps from V into $\mathbb{C}$. The natural notion of addition and scalar multiplication of linear maps gives V* a natural vector space structure.

Now, let $\mathcal{H}$ be a Hilbert space. A key result in the theory of Hilbert spaces is the Riesz lemma (see, e.g., Reed and Simon (1980)), which states that every element of $\mathcal{H}^*$ can be uniquely expressed in the form $<\psi, \cdot>$ for some $\psi \in \mathcal{H}$, i.e. every bounded linear map from $\mathcal{H}$ into $\mathbb{C}$ is of the form of "take the inner product with some fixed vector in $\mathcal{H}$." The Riesz lemma thereby provides a one-to-one, onto, antilinear correspondence between $\mathcal{H}$ and $\mathcal{H}^*$. Consequently, we obtain a natural isomorphism (i.e., a one-to-one, onto, linear map) between $\mathcal{H}^*$ and the complex conjugate space $\overline{\mathcal{H}}$. Similarly, $\mathcal{H}$ and $\overline{\mathcal{H}}^*$ are naturally isomorphic. The Riesz lemma also guarantees that if $A:\mathcal{H}_1 \to \mathcal{H}_2$ is a bounded linear map between Hilbert spaces $\mathcal{H}_1$ and $\mathcal{H}_2$, then there exists a bounded linear map $A^\dagger:\mathcal{H}_2 \to \mathcal{H}_1$, called the *adjoint* of A, such that

$$<\psi_2, A\psi_1>_2 = <A^\dagger\psi_2, \psi_1>_1 \tag{A.2.3}$$

for all $\psi_1 \in \mathcal{H}_1$, $\psi_2 \in \mathcal{H}_2$.

Next, we define the direct sum of Hilbert spaces. Let $\{\mathcal{H}_\alpha\}$ be an arbitrary collection of Hilbert spaces, indexed by α. (We will be interested only with the case where there are at most a countable number of Hilbert spaces, but no such restriction need be made for this construction.) The elements of the Cartesian product, $\times_\alpha \mathcal{H}_\alpha$, consist of collections of vectors $\{\psi_\alpha\}$ with each $\psi_\alpha \in \mathcal{H}_\alpha$. Consider, now, the subset, $V \subset \times_\alpha \mathcal{H}_\alpha$, composed of elements for which all but finitely many of the ψ_α vanish. Then V has the natural structure of an inner product space. We define the *direct sum Hilbert space*—denoted $\oplus_\alpha \mathcal{H}_\alpha$—to be the Hilbert space completion of V. It follows that in the case of a countably infinite collection of Hilbert spaces $\{\mathcal{H}_i\}$, each $\Psi \in \oplus_i \mathcal{H}_i$ consists of arbitrary sequences $\{\psi_i\}$ such that each $\psi_i \in \mathcal{H}_i$ and $\sum_i \|\psi_i\|_i^2 < \infty$.

The tensor product, $\mathcal{H}_1 \otimes \mathcal{H}_2$, of two Hilbert spaces, $\mathcal{H}_1$ and $\mathcal{H}_2$, may be defined as follows. Let V denote the set of linear maps $A:\overline{\mathcal{H}}_1 \to \mathcal{H}_2$ which have finite rank, i.e., such that the range of A is a finite dimensional subspace of $\mathcal{H}_2$. Then V has a natural vector space structure. Define an inner product on V by

$$<A,B>_V = \mathrm{tr}(A^\dagger B) \tag{A.2.4}$$

(The right side of eq. (A.2.4) is well defined, since $A^\dagger B:\overline{\mathcal{H}}_1 \to \overline{\mathcal{H}}_1$ has finite rank.) We define $\mathcal{H}_1 \otimes \mathcal{H}_2$ to be the Hilbert space completion of V. It follows that $\mathcal{H}_1 \otimes \mathcal{H}_2$ consists of all linear maps, $A:\overline{\mathcal{H}}_1 \to \mathcal{H}_2$, which satisfy the Hilbert-Schmidt condition, $\mathrm{tr}\, A^\dagger A < \infty$. Equivalently, by the Riesz lemma, $\mathcal{H}_1 \otimes \mathcal{H}_2$ consists of all bilinear maps $\alpha:\overline{\mathcal{H}}_1 \times \overline{\mathcal{H}}_2 \to \mathbb{C}$ such that

$$\sum |\alpha(\bar{e}_{1i}, \bar{e}_{2j})|^2 < \infty \tag{A.2.5}$$

where $\{\bar{e}_{1i}\}$ and $\{\bar{e}_{2i}\}$ are orthonormal bases of $\bar{\mathcal{H}}_1$ and $\bar{\mathcal{H}}_2$, respectively.

By induction, the above construction can be extended to define the tensor product, $\mathcal{H}_1 \otimes \cdots \otimes \mathcal{H}_n$, of finitely many Hilbert spaces, $\mathcal{H}_1, \ldots, \mathcal{H}_n$. It follows that $\mathcal{H}_1 \otimes \cdots \otimes \mathcal{H}_n$ consists of all multilinear maps $\alpha: \bar{\mathcal{H}}_1 \times \cdots \times \bar{\mathcal{H}}_n \to \mathbb{C}$ satisfying

$$\sum |\alpha(\bar{e}_{1i_1}, \ldots, \bar{e}_{ni_n})|^2 < \infty \qquad \text{(A.2.6)}$$

If $\mathcal{H}_1 = \mathcal{H}_2 = \cdots = \mathcal{H}_n = \mathcal{H}$, we define the symmetrized tensor product—denoted $\overset{n}{\otimes}_s \mathcal{H}$—to be the subspace of the n-fold tensor product space, $\overset{n}{\otimes} \mathcal{H}$, consisting of the maps, α, which are totally symmetric in the n-variables. The antisymmetrized tensor product—denoted $\overset{n}{\otimes}_a \mathcal{H}$—is defined similarly. A notion of tensor product applicable to infinitely many Hilbert spaces has been given by von Neumann (1938).

An important application of the above direct sum and tensor product constructions is the definition of Fock spaces. Let $\mathcal{H}$ be a Hilbert space. Then the *Fock space* associated with $\mathcal{H}$ is defined to be the Hilbert space

$$\mathcal{F}(\mathcal{H}) = \bigoplus_{n=0}^{\infty} (\overset{n}{\otimes} \mathcal{H}) \qquad \text{(A.2.7)}$$

where we define $\overset{0}{\otimes} \mathcal{H} = \mathbb{C}$. The *symmetric Fock space*, $\mathcal{F}_s(\mathcal{H})$, and *antisymmetric Fock space*, $\mathcal{F}_a(\mathcal{H})$, associated with $\mathcal{H}$ are defined by

$$\mathcal{F}_s(\mathcal{H}) = \bigoplus_{n=0}^{\infty} (\overset{n}{\otimes}_s \mathcal{H}) \qquad \text{(A.2.8)}$$

$$\mathcal{F}_a(\mathcal{H}) = \bigoplus_{n=0}^{\infty} (\overset{n}{\otimes}_a \mathcal{H}) \qquad \text{(A.2.9)}$$

A.3 Index Notation

In analyses involving tensors over a finite dimensional vector space, it is very useful at times to employ an abstract index notation. The same is true for analyses involving tensors over an infinite dimensional Hilbert space (Geroch, unpublished) and the purpose of this section is to introduce this notation. It is assumed that the reader is familiar with the abstract index notation in the finite dimensional case (see, e.g., Wald (1984a)).

Given a Hilbert space $\mathcal{H}$, we can construct the associated Hilbert spaces $\overline{\mathcal{H}}$, $\mathcal{H}^*$, and $\overline{\mathcal{H}}^*$ as in the previous section. In analogy with the index notation commonly employed for spinors (see, e.g., Wald (1984a)), it is natural to use an index notation wherein elements of $\mathcal{H}$ are denoted as ψ^a, elements of $\overline{\mathcal{H}}$ are denoted as $\phi^{a'}$, elements of $\mathcal{H}^*$ are denoted as λ_a, and elements of $\overline{\mathcal{H}}^*$ are denoted as $\xi_{a'}$. However, as noted above, using the Riesz lemma, we may identify $\overline{\mathcal{H}}$ with $\mathcal{H}^*$ and we may identify $\mathcal{H}$ with $\overline{\mathcal{H}}^*$. We shall make these identifications, and thereby eliminate the need for the introduction of primed indices. Thus, in particular, if $\psi^a \in \mathcal{H}$, we shall denote the corresponding element of $\overline{\mathcal{H}}$ as $\overline{\psi}_a$.

In analogy with the finite dimensional case, an element $\psi \in \overset{n}{\otimes}\mathcal{H}$ will be denoted as $\psi^{a_1 \cdots a_n}$. The subspace $\overset{n}{\otimes}_s\mathcal{H}$ then consists of those elements satisfying

$$\psi^{a_1 \ldots a_n} = \psi^{(a_1 \ldots a_n)} \qquad \text{(A.3.1)}$$

where the round brackets denote symmetrization. An element $\phi \in \overset{n}{\otimes}\overline{\mathcal{H}}$ will be denoted as $\phi_{a_1 \cdots a_n}$. Similarly, If, for example, $\chi \in \mathcal{H} \otimes \mathcal{H}^*$, then χ will be denoted as $\chi^a{}_b$. Outer products and contractions of tensors over $\mathcal{H}$ are denoted as in the finite dimensional case. Thus, in particular, the inner product of vectors $\phi, \psi \in \mathcal{H}$ may be denoted as

$$<\phi,\psi> = \overline{\phi}_a \psi^a \qquad \text{(A.3.2)}$$

It should be noted that in the finite dimensional case, $\mathcal{H} \otimes \mathcal{H}^*$ is isomorphic to the space of linear maps from $\mathcal{H}$ into $\mathcal{H}$, so in that case any linear map $A:\mathcal{H} \to \mathcal{H}$ can be represented in the index notation as $A^a{}_b$. However, in the infinite dimensional case, only the Hilbert-Schmidt linear maps (satisfying $\mathrm{tr} A^\dagger A < \infty$) define tensors, so index notation will be used only for such Hilbert-Schmidt maps.

An important difference occurring in the infinite dimensional case is that contractions over an upper and lower index involve infinite sums and thus need not always exist. In particular, even if $A:\mathcal{H} \to \mathcal{H}$ is Hilbert-Schmidt—and thus defines a tensor, $A^a{}_b$—its trace need not exist. However, the condition (A.2.6) guarantees that compositions of tensors—i.e., contractions over upper and lower indices occurring on different tensors—always is well defined. Thus, for example, although $A^a{}_a$ need not exist, a contraction of the form $A^a{}_{bc}B^{bd}$ or $A^a{}_{bc}B^{bc}$ always is well defined for any $A \in \mathcal{H} \otimes \mathcal{H}^* \otimes \mathcal{H}^*$, $B \in \mathcal{H} \otimes \mathcal{H}$.

We conclude this Appendix by giving the general definition of annihilation and creation operators on a symmetric Fock space, thereby providing a good illustration of the use of the index notation. Let $\mathcal{H}$ be a Hilbert space and let $\mathcal{F}_s(\mathcal{H})$ be its associated symmetric Fock space, as defined at the end of the previous section. In the index notation, a vector $\Psi \in \mathcal{F}_s(\mathcal{H})$ can be represented as

$$\Psi = (\psi, \psi^{a_1}, \psi^{a_1 a_2}, \ldots, \psi^{a_1 \ldots a_n}, \ldots) \qquad \text{(A.3.3)}$$

where, for all n, we have $\psi^{a_1 \ldots a_n} = \psi^{(a_1 \ldots a_n)}$. Now, let $\xi^a \in \mathcal{H}$ and let $\bar{\xi}_a$ denote the corresponding element of $\bar{\mathcal{H}}$. The annihilation operator $a(\bar{\xi}): \mathcal{F}_s(\mathcal{H}) \to \mathcal{F}_s(\mathcal{H})$ associated with $\bar{\xi}$ is defined by

$$a(\bar{\xi})\Psi = (\bar{\xi}_a \psi^a, \sqrt{2}\, \bar{\xi}_a \psi^{a a_1}, \sqrt{3}\, \bar{\xi}_a \psi^{a a_1 a_2}, \ldots) \qquad \text{(A.3.4)}$$

(Note that we chose to view the annihilation operator as a function of vectors in $\bar{\mathcal{H}}$ rather than $\mathcal{H}$ in order to make its dependence upon these vectors be linear rather than antilinear. Note also that we dropped the index on $\bar{\xi}_a$ when writing $a(\bar{\xi})$ since its presence in that expression could cause confusion.) Similarly, the creation operator $a^\dagger(\xi): \mathcal{F}_s(\mathcal{H}) \to \mathcal{F}_s(\mathcal{H})$ associated with ξ^a is defined by

$$a^\dagger(\xi)\Psi = (0, \psi\xi^{a_1}, \sqrt{2}\, \xi^{(a_1}\psi^{a_2)}, \sqrt{3}\, \xi^{(a_1}\psi^{a_2 a_3)}, \ldots) \qquad \text{(A.3.5)}$$

If the domains of $a(\bar{\xi})$ and $a^\dagger(\xi)$ are defined to be the subspaces of $\mathcal{F}_s(\mathcal{H})$ such that the norms of the right sides of eqs. (A.3.4) and (A.3.5), respectively, are finite, then it may be verified that $a^\dagger(\xi)$ is, indeed, the adjoint of $a(\bar{\xi})$ (see, e.g., Reed and Simon (1980) for the definition of the adjoint of an unbounded operator). It also may be readily verified that $a(\bar{\xi})$ and $a^\dagger(\eta)$ satisfy the commutation relation,

$$[a(\bar{\xi}), a^\dagger(\eta)] = \bar{\xi}_a \eta^a\, I \qquad \text{(A.3.6)}$$

References

Allen, B. (1985), "Vacuum States in deSitter Space", Phys. Rev. **D32**, 3136.

Arnold, V.I. (1989), *Mathematical Methods of Classical Mechanics* (Springer-Verlag, New York).

Ashtekar, A. (1991), *Lectures on Non-Perturbative Canonical Gravity* (World Scientific Press, Singapore).

Ashtekar, A., and Magnon, A. (1975), "Quantum Fields in Curved Space-Times", Proc. Roy. Soc. Lond. **A346**, 375.

Banks, T. Susskind, L., and Peskin, M.E. (1984), "Difficulties for the Evolution of Pure States into Mixed States", Nucl. Phys. **B244**, 125.

Bardeen, J.M., Carter, B., and Hawking, S.W. (1973), "The Four Laws of Black Hole Mechanics", Commun. Math. Phys. **31**, 161.

Bekenstein, J.D. (1974), "Generalized Second Law of Thermodynamics in Black Hole Physics", Phys. Rev. **D9**, 3292.

Bekenstein, J.D. (1981), "Universal Upper Bound on the Entropy-to-Energy Ratio for Bounded Systems", Phys. Rev. **D23**, 287.

Bombelli, L., Koul, R.K., Lee, J., and Sorkin, R.D.(1986), "Quantum Source of Entropy for Black Holes", Phys. Rev. **D34**, 373.

Birrell, N.D., and Davies, P.C.W. (1982), *Quantum Fields in Curved Space* (Cambridge University Press, Cambridge)

Bisognano, J.J., and Wichmann, E.H. (1976), "On the Duality Condition for Quantum Fields", J. Math. Phys. **17**, 303.

Braden, H.W., Whiting, B.F., and York, J.W. (1987), "Density of States for the Gravitational Field in Black Hole Topologies", Phys. Rev. **D36**, 3614.

Bratteli, O., and Robinson, D.W. (1981) *Operator Algebras and Quantum Statistical Mechanics II* (Springer-Verlag, New York).

Brown, J.D., Comer, G.L., Martinez, E.A., Melmed, J., Whiting, B.F., and York, J.W. (1990), "Thermodynamic Ensembles and Gravitation", Class. Quantum Grav. **7**, 1433.

Brown, J.D., and York, J.W. (1993a), "Quasilocal Energy and Conserved Charges Derived from the Gravitational Action", Phys. Rev. **D47**, 1407.

Brown, J.D., and York, J.W. (1993b), "Microcanonical Functional Integral for the Gravitational Field", Phys. Rev. **D47**, 1420.

Callen, C.G., Giddings, S.B, Harvey, J.A., and Strominger, A. (1992), "Evanescent Black Holes", Phys. Rev. **D45**, 1005.

Chernoff, P.R. (1981), "Mathematical Obstructions to Quantization", Hadronic Journal **4**, 879.

Chmielowski, P. (1994), "States of a Scalar Field on Spacetimes with Two Isometries with Timelike Orbits", Class. Quant. Grav. **11**, 41.

Christodoulou, D. (1994), "Examples of Naked Singularity Formation in the Gravitational Collapse of a Scalar Field", Ann. Math. (in press).

DeWitt, B.S. (1975), "Quantum Field Theory in Curved Spacetime", Phys. Rep. **19**, 295.

Dieckmann, J. (1988), "Cauchy Surfaces in Globally Hyperbolic Spacetimes", J. Math. Phys. **29**, 578.

Dimock, J. (1980), "Algebras of Local Observables on a Manifold", Commun. Math. Phys. **77**, 219.

Dimock, J. (1982), "Dirac Quantum Fields on a Manifold", Trans. Amer. Math. Soc. **269**, 133.

Dimock, J. (1985), "Scattering for the Wave Equation on the Schwarzschild Metric", Gen. Rel. and Grav. **17**, 353.

Dimock, J. (1992), "Quantized Electromagnetic Field on a Manifold", Rev. Math. Phys. **4**, 223.

Dimock, J., and Kay, B.S. (1987), "Classical and Quantum Scattering Theory for Linear Scalar Fields on the Schwarzschild Metric. I", Ann. Phys. **175**, 366.

Fell, J.M. (1960), "The Dual Spaces of C*-Algebras", Trans. Am. Math. Soc. **94**, 365.

Fredenhagen, K., and Haag, R. (1990), "On the Derivation of the Hawking Radiation Associated with the Formation of a Black Hole", Commun. Math. Phys. **127**, 273.

Frolov, V.P. (1992), "Two-Dimensional Black Hole Physics", Phys. Rev. **D46**, 5383.

Frolov, V.P., and Page, D.N. (1993), "Proof of the Generalized Second Law for Quasistationary Semiclassical Black Holes", Phys. Rev. Lett. **71**, 3902.

Fulling, S.A. (1973), "Nonuniqueness of Canonical Field Quantization in Riemannian Space-Time", Phys Rev. **D7**, 2850.

Fulling, S.A. (1989), *Aspects of Quantum Field Theory in Curved Spacetime* (Cambridge University Press, Cambridge).

Fulling, S.A., Narcowich, F., and Wald, R.M. (1981), "Singularity Structure of the Two-Point Function in Quantum Field Theory in Curved Spacetime. II", Ann. Phys. **136**, 243.

Fulling, S.A., and Ruijsenaars (1987), "Temperature, Periodicity, and Horizons", Phys. Rep. **152**, 135.

Fulling, S.A., Sweeny, M., and Wald, R.M. (1978), "Singularity Structure of the Two-Point Function in Quantum Field Theory in Curved Spacetime", Commun. Math. Phys. **63**, 257.

Garabedian, P.R. (1964), *Partial Differential Equations* (Wiley, New York).

Geroch, R. (1970), "Domain of Dependence", J. Math. Phys. **11**, 437.

Gibbons, G.W., and Hawking, S.W. (1977a), "Cosmological Event Horizons, Thermodynamics, and Particle Creation", Phys. Rev. **D15**, 2738.

Gibbons, G.W., and Hawking, S.W. (1977b), "Action Integrals and Partitiion Functions in Quantum Gravity", Phys. Rev. **D15**, 2752.

Gibbons, G.W., and Perry, M.J. (1978), "Black Holes and Thermal Green's Functions", Proc. Roy. Soc. Lond. **A358**, 467.

Gotay, M.J. (1980), "Functorial Geometric Quantization and Van Hove's Theorem", Int. J. Theor. Phys. **19**, 139.

Haag, R. (1992), *Local Quantum Physics* (Springer-Verlag, Berlin).

Haag, R., and Kastler, D. (1964), "An Algebraic Approach to Quantum Field Theory", J. Math. Phys. **5**, 848.

Hadamard, J. (1923), *Lectures on Cauchy's Problem in Linear Partial Differential Equations* (Yale University Press, New Haven).

Hartle, J.B., and Hawking, S.W. (1976), "Path-Integral Derivation of Black Hole Radiance", Phys. Rev. **D13**, 2188.

Hawking, S. (1971), "Gravitational Radiation From Colliding Black Holes", Phys. Rev. Lett. **26**, 1344.

Hawking, S.W. (1975), "Particle Creation by Black Holes", Commun. Math. Phys. **43**, 199.

Hawking, S.W. (1976), "Breakdown of Predictability in Gravitational Collapse", Phys. Rev. **D14**, 2460.

Hawking, S.W., and Ellis. G.F.R. (1973), *The Large Scale Structure of Space-Time* (Cambridge University Press, Cambridge).

Hawking, S.W., and Hartle, J. (1972), "Energy and Angular Momentum Flow into a Black Hole", Commun. Math. Phys. **27**, 283.

Huang, K. (1963), *Statistical Mechanics* (Wiley , New York).

Isham, C.J. (1984), "Topological and Global Aspects of Quantum Theory", in *Relativity, Groups, and Topology II,* ed. B.S. DeWitt and R. Stora (North-Holland, Amsterdam).

Israel, W. (1976), "Thermo-Field Dynamics of Black Holes", Phys. Lett. **57A**, 107.

Israel, W. (1986), "Third Law of Black Hole Dynamics: A Formulation and Proof", Phys. Rev. Lett. **57**, 397.

Iyer, V. and Wald, R.M. (1994), "Some Properties of Noether Charge and a Proposal for Dynamical Black Hole Entropy", Phys. Rev. D (in press).

Jacobson, T.A. (1993), "Black Hole Radiation in the Presence of a Short Distance Cutoff", Phys. Rev. **D48**, 728.

Jacobson, T.A., and Kang, G. (1993), "Conformal Invariance of Black Hole Temperature", Class. Quant. Grav. **10**, L1.

Kay, B.S. (1978), "Linear Spin-Zero Quantum Fields in External Gravitational and Scalar Fields", Commun. Math. Phys. **62**, 55.

Kay, B.S. (1985), "The Double-Wedge Algebra for Quantum Fields on Schwarzschild and Minkowski Spacetimes", Commun. Math. Phys. **100**, 57.

Kay, B.S. (1988), "Quantum Field Theory in Curved Spacetime", in *Differential Geometrical Methods in Theoretical Physics,* ed. K. Bleuler and M. Werner (Reidel, Dordrecht).

Kay, B.S. (1992), "The Principle of Locality and Quantum Field Theory on (Non Globally Hyperbolic) Curved Spacetimes", Reviews in Mathematical Physics (Special Issue), 167.

Kay, B.S. (1993), "Sufficient Conditions for Quasifree States and an Improved Uniqueness Theorem for Quantum Fields on Spacetimes with Horizons", J. Math. Phys. **34**, 4519.

Kay, B.S., and Wald, R.M. (1991), "Theorems on the Uniqueness and Thermal Properties of Stationary, Nonsingular, Quasifree States on Spacetimes with a Bifurcate Killing Horizon", Phys. Rep. **207**, 49.

Lansford, O.E. (1971), "Selected Topics in Functional Analysis", in *Statistical Mechanics and Quantum Field Theory,* ed. C. DeWitt and R. Stora (Gordon and Breach, New York).

Lee, J., and Wald, R.M. (1990), "Local Symmetries and Constraints", J. Math. Phys. **31**, 725.

Parker, L. (1969), "Quantized Fields and Particle Creation in Expanding Universes", Phys. Rev. **183**, 1057.

Racz, I., and Wald, R.M. (1992), "Extensions of Spacetimes with Killing Horizons", Class. Quant. Grav. **9**, 2643.

Radzikowski, M. (1992), "The Hadamard Condition and Kay's Conjecture in (Axiomatic) Quantum Field Theory in Curved Spacetime", Ph. D. thesis, Princeton University.

Reed, M., and Simon, B. (1980), *Functional Analysis* (Academic Press, London).

Russo, J.G., Susskind, L., and Thorlacius, L. (1992), "End Point of Hawking Radiation", Phys. Rev. **D46**, 3444.

Schoen, R., and Yau, S.-T. (1983), "The Existence of a Black Hole due to Condensation of Matter", Commun. Math, Phys. **90**, 575.

Segal, I.E. (1963), *Mathematical Problems of Relativistic Physics* (Am. Math. Soc., Providence).

Segal, I.E. (1967), "Representations of the Canonical Commutation Relations", in *Applications of Mathematics to Problems in Theoretical Physics,* ed. F. Lurcat (Gordon and Breach, New York).

Sewell, G.L. (1982), "Quantum Fields on Manifolds: PCT and Gravitationally Induced Thermal States", Ann. Phys. **141**, 201.

Shale, D. (1962), "Linear Symmetries of Free Boson Fields", Trans. Am. Math. Soc. **103**, 149.

Simon, B. (1972), "Topics in Functional Analysis", in *Mathematics of Contemporary Physics,* ed. R.F. Streater (Academic Press, London).

Simon, J.Z. (1990), "Higher Derivative Lagrangians, Nonlocality, Problems, and Solutions", Phys. Rev. **D41**, 3720.

Slawny, J. (1972), "On Factor Representations and the C*-Algebra of the Canonical Commutation Relations", Commun. Math. Phys. **24**, 151.

Streater, R.F., and Wightman, A.S. (1964), *PCT, Spin and Statistics, and All That* (Benjamin, New York).

Sudarsky, D., and Wald, R.M. (1992), "Extrema of Mass, Stationarity and Staticity, and Solutions to the Einstein-Yang-Mills Equations", Phys. Rev. **D46**, 1453.

't Hooft, G. (1991), "The Black Hole Horizon as a Quantum Surface", Physica Scripta **T36**, 247.

't Hooft, G. (1992), "Scattering Matrix for a Quantized Black Hole", in *Black Hole Physics.*, ed. V. DeSabbata and Z. Zhang, (Kluwer Academic Publishers, Dordrecht).

Tagaki, S. (1986), "Vacuum Noise and Stress Induced by Uniform Acceleration", Prog. Theor. Phys. Suppl. **88**, 1.

Thorne, K.S., Zurek, W.H., and Price, R.H. (1986), "The Thermal Atmosphere of a Black Hole", in *Black Holes: The Membrane Paradigm,* ed. K. S. Thorne, R. H. Price, and D. A. MacDonald (Yale University Press, New Haven).

Unruh, W.G. (1976), "Notes on Black Hole Evaporation", Phys. Rev. **D14**, 870.

Unruh, W.G., and Wald, R.M. (1984), "What Happens When an Accelerating Observer Detects a Rindler Particle", Phys. Rev. **D29**, 1047.

Verch, R. (1994), "Local Definiteness, Primarity, and Quasiequivalence of Quasi-free Hadamard Quantum States in Curved Spacetime", Commun. Math. Phys. **160**, 507.

von Neumann, J. (1938), "On Infinite Direct Products", Composito Mathematica **6**, 1.

Wald, R.M. (1975), "On Particle Creation by Black Holes", Commun. Math. Phys. **45**, 9.

Wald, R.M. (1976), "Stimulated Emission Effects in Particle Creation near Black Holes", Phys. Rev. **D13**, 3176.

Wald, R.M. (1977), "The Back Reaction Effect in Particle Creation in Curved Spacetime", Commun. Math. Phys. **54**, 1.

Wald, R.M. (1978a), "Trace Anomaly of a Conformally Invariant Quantum Field in Curved Spacetime", Phys. Rev. **D17**, 1477.

Wald, R.M. (1978b), "Axiomatic Renormalization of the Stress Tensor of a Conformally Invariant Field in Conformally Flat Spacetimes", Ann. Phys. **110**, 472.

Wald, R.M. (1979a), "Existence of the S-Matrix in Quantum Field Theory in Curved Spacetime", Ann. Phys. **118**, 490.

Wald, R.M. (1979b), "On the Euclidean Approach to Quantum Field Theory in Curved Spacetime", Commun. Math. Phys. **70**, 221.

Wald, R.M. (1980), "Quantum Gravity and Time Reversibility", Phys. Rev. **D21**, 2742.

Wald, R.M. (1984a), *General Relativity* (University of Chicago Press, Chicago).

Wald, R.M. (1984b), "Black Holes, Singularities, and Predictability", in *Quantum Theory of Gravity*, ed. S.M. Christensen (Adam Hilger Press, Bristol).

Wald, R.M. (1986), "Black Holes and Quantum Coherence", Found. of Phys. **16**, 501.

Wald, R.M. (1988), "Black Hole Thermodynamics", in *Highlights in Gravitation and Cosmology*, ed. B. R. Iyer, A. Kembhavi, J. V. Narlikar, and C. V. Vishveshwara, (Cambridge University Press, Cambridge).

Wald, R.M. (1992), "Black Holes and Thermodynamics", in *Black Hole Physics.*, ed. V. DeSabbata and Z. Zhang, (Kluwer Academic Publishers, Dordrecht)

Wald, R.M. (1993a), "The First Law of Black Hole Mechanics", in *Directions in General Relativity,* Vol.1, ed. B.L. Hu, M.P. Ryan, and C.V. Vishveshwara, (Cambridge University Press, Cambridge).

Wald, R.M. (1993b), "'Weak' Cosmic Censorship", in *PSA 1992*, volume 2, ed. by M. Forbes D. Hull and K. Okruhlik, Philosophy of Science Association (East Lansing, 1993).

Wald, R.M. (1993c), "Black Hole Entropy is the Noether Charge", Phys. Rev. **D48**, R3427.

Woodhouse, N. (1980) *Geometric Quantization* (Clarendon Press, Oxford).

Zurek, W.H., and Thorne, K.S. (1986), "Statistical Mechanical Origin of the Entropy of a Rotating, Charged Black Hole", Phys. Rev. Lett. **54**, 2171.

Notation Index

General Index

Italic page numbers denote definitions.